UFO & ALIEN COLLECTIBLES

PRICE GUIDE

DANA CAIN

Published by

700 E. State Street • Iola, WI 54990-0001
Telephone: 715/445-2214

Please, call or write us for our free catalog of antiques and collectibles publications.
To place an order or receive our free catalog, call 800-258-0929.
For editorial comment and further information,
use our regular business telephone at (715) 445-2214.

ISBN: 0-87341-690-2

Printed in the United States of America

Dedication

This one is for
The Denver Area Science Fiction Association
founded in the late 1960s and still a hot spot for
speculation, wisdom, friendship and fun and for its 1997-98 director
Gail Barton
without question one of the most far out, paranormal, intelligent,
imaginative, talented, weird, creative and spectacular people I have
ever or will ever have the pleasure to know.

Acknowledgments

Alien Spacecraft? UFO Research Center • Roswell, New Mexico
Alien Zone • Roswell, New Mexico
Atomic Antiques • Denver, Colorado
Gail Barton • Lakewood, Colorado (book and magazine collection)
Rose Beetem • Denver, Colorado (X-Files collection)
Bruce Carteron - Cinema Graphics • Denver, Colorado (movie posters)
Brian Cooper • Hollywood, California
Kent Cordray • Denver, Colorado
Charles Hickson • Gautier, Mississippi
Thea Hutcheson • Sheridan, Colorado
International UFO Museum & Research Center • Roswell, New Mexico
Robert Klippel • Denver, Colorado
Southwest Alien • Roswell, New Mexico
Space Station • Denver, Colorado
Linda Stanley • Fort Collins, Colorado (E.T. collection)
Starchild • Roswell, New Mexico

Contents

Scope of this Book **6**

Introduction **7**

Chapter 1: Space Toys Then and Now **8**

Special Focus: The Pascagoula UFO—An Interview with Charles Hickson **21**

Chapter 2: Books, Periodicals and Other Sacred Texts **23**

Chapter 3: UFOs and Aliens in the Movies **58**

Special Focus: E.T. the Extraterrestrial **76**

Chapter 4: It Came From the TV Set **113**

Special Focus: The X-Files. **123**

Chapter 5: Aliens at Home. **138**

Chapter 6: Roswell Mania and Area 51 . . . . **146**

Scope of this Book

Mr. Spock is not in this book. *Star Wars* is not in this book. Our focus here is on aliens coming to earth. So, we cover *Alf*, but not *Alien*. The TV and Movie chapters may seem to be sporadic, because we have tried to maintain this focus, and also showcase some of the shows which feature a variety of collectibles. *Mars Needs Women* is a great film, for example, but, unfortunately, no collectibles were produced. At any rate, we hope your favorites are well represented.

Introduction

We are not alone. Watch the skies. The truth is out there.

A lot has changed in the past 50 years. Since the big flying saucer flap of the late 1940s, more and more people have begun to accept the notion that there may be other forms of intelligent life in the universe. More and more people believe that it may even be possible that some of those life forms have made contact with Earth.

More and more people believe we are not alone.

We are drawn to the notion of contacting alien beings...someone who knows more than we do, someone who represents the incredible destiny our own human evolution may eventually approach. Will we one day have big heads, telepathic powers and interstellar transportation?

Do these aliens hold the secrets of our own future? Are they affecting it now?

Perhaps they also speak to our fear of the unknown. They are not like us. We don't know what their intentions are, but we're pretty sure they can hurt us. They represent a threat on a global level and on a personal psychic level.

Are they really here, walking among us, or did we invent them as a concrete representation of our greatest fears and our greatest hopes?

Whether or not aliens are real, they have made a huge impact on human culture. They have challenged and changed the way we view ourselves, our culture, our future and our universe.

No wonder we find the image so compelling. An alien form or face, a flying saucer, a glowing UFO, all serve as reminders that maybe we need to be more humble, more conscious, more advanced, more aware. Maybe we need to look forward more, pay more attention to things beyond our everyday experience.

Maybe there's much more to know. Maybe we're not alone.

And so, we're drawn to the images. Collectors of UFO and alien memorabilia often focus first and foremost on the information, the books and periodicals which attempt to explain the phenomenon that has swept the globe. UFOs have been seen in nearly every part of the world. Collectors want to know about the latest research, the latest sightings, the latest abduction reports, the latest rumors.

And in many ways, that is the heart of this hobby. We have been told "The truth is out there," and we want to find it.

Chapter 1

Space Toys Then and Now

Toys based on the notion of flying saucers and alien invaders first became popular in the 1950s, a few years after the initial flying saucer "scare" gripped the nation. These toys, which are now quite collectible, paved the way for many successful UFO and alien toy lines in the following decades. The late 1990s has seen a fantastic increase in the number of alien-related toys, thanks to the public's fascination with Roswell and the X-Files notion of alien visitors.

We can see how our ideas about aliens have changed over the decades by looking at toy design. From the Waxy Miller Alien figures of the late 1950s to Colorforms' Outer Space Men of the 1960s and Tomland's alien figures of the 1970s, the image has evolved. Whereas earlier alien figures were based on science fiction, speculation and imagination, today's alien toys are based primarily on reported sightings and thousands of abductee descriptions. The image of the classic "Grey" alien is known to all and has become a classic format for modern-day action figures, dolls and toys.

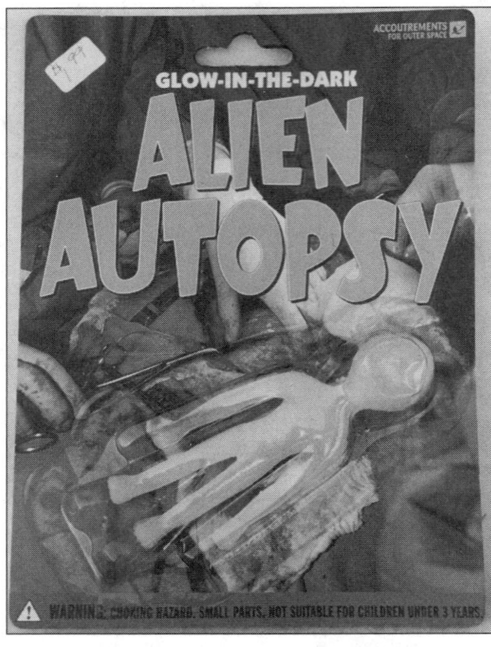

Alien Autopsy figure, glows, 3-1/2", carded, Accoutrements, 1996-$3.

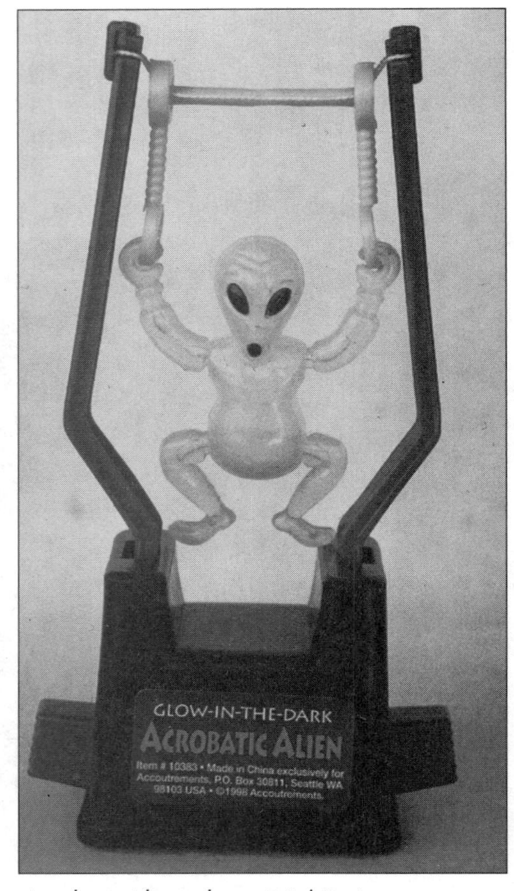

Acrobatic Alien, glows, 5-1/4", Accoutrements, 1998-$4.

Action figures, 4", variety of aliens, Tomland, 1976, each-$10, MIP.

Item	Good/Loose	Mint/MIP
Acrobatic Alien, glows, 5-1/4", Accoutrements, 1998	2.00	4.00
Action figures, 4", variety of aliens, Tomland, 1976, each	5.00	10.00
Alien Autopsy figure, glows, 3-1/2", carded, Accoutrements, 1996	1.00	3.00
Alien Dissection Doll, with organs inside, green rubber, 14-1/2", R. Marino, WPF Inc., 1997	12.00	18.00
Alien Lifeform figure, 8", Shadow Box Collectibles, 1996	9.00	14.00
Alien Lifeform with Panoramic Display, 7", Shadow Box Collectibles, 1995	8.00	12.00
Alien Series, action figure, Grey Alien with trading card, Shadow Box, 1997	5.00	7.00
Aliens...The Mini Series, Ancient Alien, 4", Shadow Box Collectibles, 1997	5.00	7.00
Aliens...The Mini Series, Chupacabra Alien, 3-1/2", Shadow Box Collectibles, 1997	5.00	7.00
Aliens...The Mini Series, Grey Alien, 4", Shadow Box Collectibles, 1997	5.00	7.00
Aliens...The Mini Series, Hybrid Alien, 4", Shadow Box Collectibles, 1997	5.00	7.00
Aliens...The Mini Series, Men in Black Alien, 5", Shadow Box Collectibles, 1997	5.00	7.00
Aliens...The Mini Series, Neonate Alien, 3-1/2", Shadow Box Collectibles, 1997	5.00	7.00
Aliens...The Mini Series, Nordic Alien, 4", Shadow Box Collectibles, 1997	5.00	7.00
Aliens...The Mini Series, Reptilian Alien, 5", Shadow Box Collectibles, 1997	5.00	7.00
Aliens...The Mini Series, Robot Alien, 5", Shadow Box Collectibles, 1997	5.00	7.00
Aliens...The Mini Series, Roswell Alien, 4", Shadow Box Collectibles, 1997	5.00	7.00

Alien Lifeform with Panoramic Display, 7", Shadow Box Collectibles, 1995-$12.

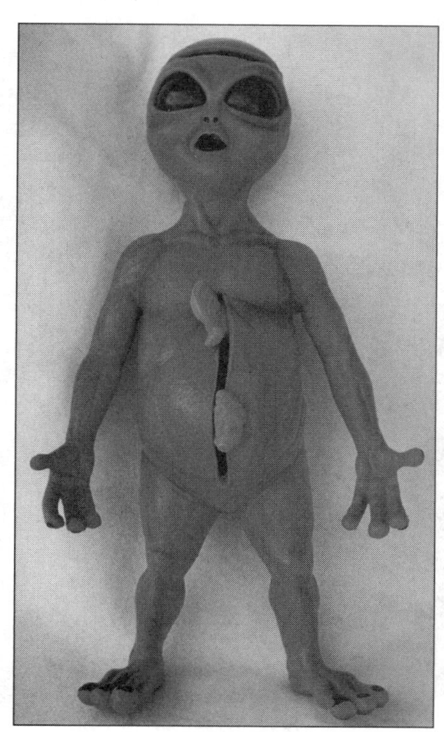

Alien Autopsy figure, glows, 3-1/2", carded, Accoutrements, 1996-$3, MIP.

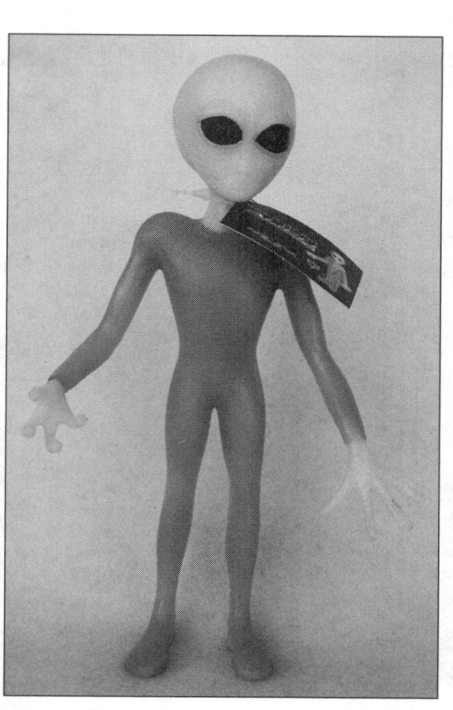

Alien Lifeform figure, 8", Shadow Box Collectibles, 1996-$14, MIP.

Alien Series, action figure, Grey Alien with trading card, Shadow Box, 1997-$7.

Item	Good/Loose	Mint/MIP
Bean bag green alien, Magnatoes, 5", Mary Meyer Corp., 1997	4.00	6.00
Big Loo–Your Friend From the Moon, 38" tall, plastic, Marx, 1960s	1,500.00	3,000.00
Buck Rogers Flying Saucer, printed paper plates with metal rim, 6" across, S.P. Co.	400.00	600.00

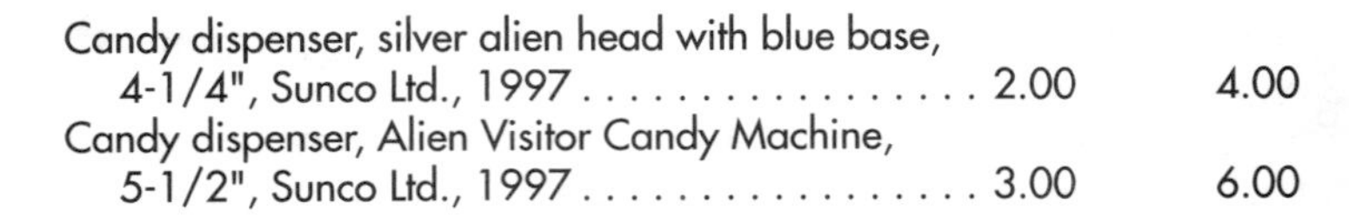

Item	Good/Loose	Mint/MIP
Candy dispenser, silver alien head with blue base, 4-1/4", Sunco Ltd., 1997	2.00	4.00
Candy dispenser, Alien Visitor Candy Machine, 5-1/2", Sunco Ltd., 1997	3.00	6.00

Aliens...The Mini Series, Ancient Alien, 4", Chupacabra Alien, 3-1/2", Grey Alien, 4", Hybrid Alien, 4", Men in Black Alien, 5", Neonate Alien, 3-1/2", Nordic Alien, 4", Reptilian Alien, 5", Robot Alien, 5", Roswell Alien, 4", Shadow Box Collectibles, 1997, each-$7, MIP.

Bean bag green alien, Magnatoes, 5", Mary Meyer Corp., 1997-$4-6.

Candy dispenser, silver alien head with blue base, 4-1/4", Sunco Ltd., 1997-$4.

Candy dispenser, Alien Visitor Candy Machine, 5-1/2", Sunco Ltd., 1997-$6.

Item	Good/Loose	Mint/MIP
Cereal premium, "Crunchy," red plastic alien, 2", 1960s(?)	8.00	15.00
Cereal premium, alien with bird-head, bronze-colored plastic, 1-3/4", 1950s	12.00	20.00
Cereal premium, robot alien, soft red plastic, 2", 1950s	12.00	20.00
Coloring book, UFO–Seeing is Believing, Whitman 1019, 1974 version	8.00	12.00
Coloring book, UFO–Space Strangers, Whitman, 1968-1978 editions	8.00	15.00

Cereal premium, robot alien, soft red plastic, 2", 1950s-$20.

Cereal premium, "Crunchy," red plastic alien, 2", 1960s(?)-$15.

Coloring book, UFO - Space Strangers, Whitman, 1968-1978 editions-$15.

Cereal premium, alien with bird-head, bronze-colored plastic, 1-3/4", 1950s-$20.

Item	Good/Loose	Mint/MIP
Figure, green vinyl alien with silver "speedo," 1990s	8.00	12.00
Finger puppet, plush alien, "My name is Cosmos," 7", Mary Meyer Tippy Toes, 1990s	4.00	7.00
Flying Saucer, battery-op toy, 9" diameter, Y Co., 1970s	100.00	175.00
Flying Saucer with Space Pilot, tin battery-op toy, 7-1/2" diameter, 1950s	350.00	500.00
Flying Space Saucer, battery-op toy (with plastic missiles), 9" diameter, A.S.C. Co., 1960s	400.00	600.00
Frisbee, Alien Freaky Flying Disc, alien face, 9", Flight Force, WPF Inc., 1997	12.00	15.00
Frisbee, Flying Saucer Flying Disc, 9", Flight Force, WPF Inc., 1997	12.00	15.00

Finger puppet, plush alien, "My name is Cosmos," 7", Mary Meyer Tippy Toes, 1990s-$7.

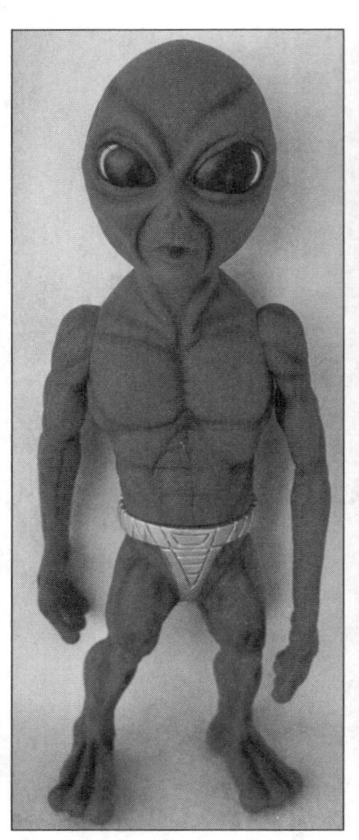
Figure, green vinyl alien with silver "speedo," 1990s-$12, MIP.

Coloring book, UFO - Space Strangers, Whitman, 1968-1978 editions-$15.

Frisbee, Alien Freaky Flying Disc, alien face, 9", Flight Force, WPF Inc., 1997-$15.

Item	Good/Loose	Mint/MIP
Game, Alien Autopsy Game, (similar to "Operation"), DaMert Company, 1997	8.00	15.00
Game, UFO Game of Close Encounters, AH Bookshelf Games, Avalon Hill, 1978	8.00	15.00
Game, UFOria, Steve Falcore, 1989	15.00	25.00
GloHomies, aliens from the 'hood, 3" tall, boxed, Vandor China, 1997	7.00	10.00
Glow in the Dark Aliens set, 4 figures, 4" to 5", Shadow Box Collectibles, 1997	8.00	12.00

Game, UFO Game of Close Encounters, AH Bookshelf Games, Avalon Hill, 1978-$15.

GloHomies, aliens from the 'hood, 3" tall, boxed, Vandor China, 1997-$10.

Game, UFOria, Steve Falcore, 1989-$25.

Glow in the Dark Aliens set, 4 figures, 4-5", Shadow Box Collectibles, 1997-$12.

Item	Good/Loose	Mint/MIP
Jupiter Flying Saucer, tin and plastic battery-op toy, 8" diameter, KO Co., 1960s	175.00	275.00
King Flying Saucer, battery-op toy, 12-1/2" diameter, KO Co., 1960s	150.00	300.00
Mask, full head latex alien mask, Fright Asylum disguise, 1997	12.00	18.00
McDonald's Space Aliens, set of 8 rubber aliens, Happy Meal toys 1979, each	3.00	4.00
Mercury X-1 Space Saucer, battery-op toy, 8" diameter, Y Co., 1960s	75.00	150.00
Model kit, Area S4 UFO, Testors #576, 1995	20.00	30.00
Model kit, "Grey," 1/6 scale, Testors #761, 1995	8.00	15.00
Model kit, UFO, classic saucer with little green alien inside, Lindberg, 1956	125.00	175.00
Model kit, UFO, glows in the dark, re-issue of Lindberg's 1956 kit, 1972	40.00	65.00
Moon Ship, battery-op toy, 8" diameter, Toy Town Tomy, Japanese, 1970s	60.00	120.00
Mystery Space Ship, 50+ tricks, 7" x 11" box, 8" diameter ship, Marx, 1960	100.00	150.00
Nodder, Martian, 7", blue vinyl, 1960s	30.00	50.00

Model kit, "Grey," 1/6 scale, Testors #761, 1995-$15.

Moon Ship, battery-op toy, 8" diameter, Toy Town Tomy, Japanese, 1970s-$120.

Mask, full head latex alien mask, Fright Asylum disguise, 1997-$18.

Item	Good/Loose	Mint/MIP
Outer Space Men, Alpha 7 (Mars), bendee, 3", carded, Colorforms, 1968	50.00	75.00
Outer Space Men, Astro-Nautilus (Neptune), bendee, 5-1/2", carded, Colorforms, 1968	100.00	150.00
Outer Space Men, Colossus Rex (Jupiter), bendee, 6", carded, Colorforms, 1968	100.00	150.00
Outer Space Men, Commander Comet (Venus), bendee, 6", carded, Colorforms, 1968	75.00	100.00
Outer Space Men, Electron+ (Pluto), bendee, 5", carded, Colorforms, 1968	75.00	100.00
Outer Space Men, Orbitron (Uranus), bendee, 6", carded, Colorforms, 1968	80.00	110.00
Outer Space Men, Xodiac (Saturn), bendee, 5", carded, Colorforms, 1968	80.00	110.00
Paddle ball, Alien High Flyer Paddle Ball game, bagged, 9", made in China	2.00	4.00
Pinwheel, black with alien head print, 12-1/2", no mark	1.00	4.00
Playing cards, "The Alien Deck," Paulson, 1990s	5.00	9.00
Plush toy, green alien, 11-1/2" tall, Oriental Trading Co., 1990s	5.00	8.00
Plush toy, with plastic head, "cosmic sounds," eyes light up, 8-1/4", Gemmy Ind. Corp., 1996	6.00	10.00
Puppet, Alien Punching puppet, set of 2 different, Accoutrements, 1990s	15.00	20.00
Push puppet, green alien, 5", Accoutrements, 1997	2.00	4.00

Playing cards, "The Alien Deck," Paulson, 1990s-$5-9.

Paddle ball, Alien High Flyer Paddle Ball game, bagged, 9", made in China-$4.

Paddle ball, Alien High Flyer Paddle Ball game, bagged, 9", made in China-$4.

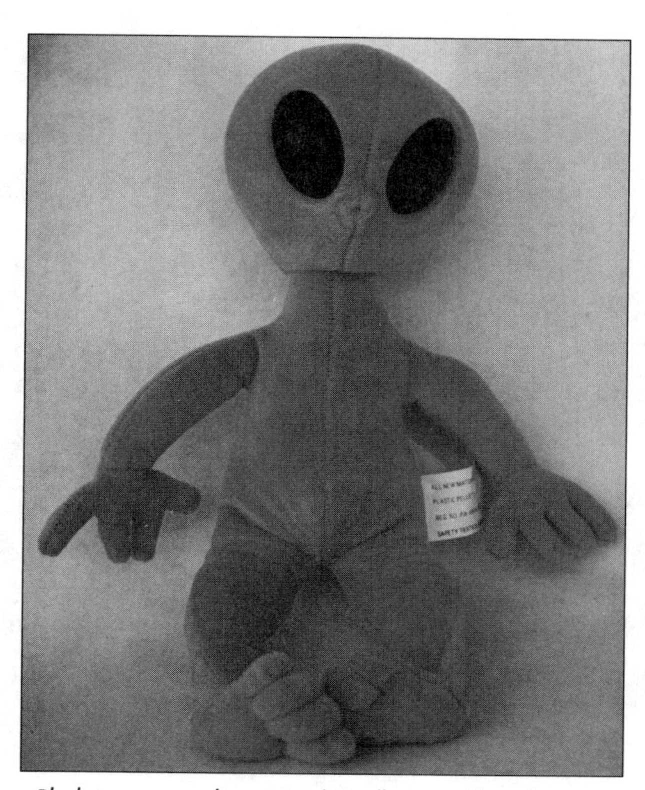
Plush toy, green alien, 11-1/2" tall, Oriental Trading Co., 1990s-$8.

Item	Good/Loose	Mint/MIP
Puzzle, Visitors, shows alien head over pyramid, 1,000 pieces, 1990s	16.00	22.00
Rex Mars Planet Patrol playset #7040, Marx, 1950s	300.00	500.00
Rex Mars Planet Patrol Target Set, tin litho base with plastic alien target figures, 15" long, Marx, 1950s	100.00	150.00
Rex Mars Space Drome playset #7016, Marx, 1954	400.00	650.00
Scorpio, purple alien, battery-op bendee, from Major Matt Mason line, 8 in., Mattel, 1969	200.00	350.00
Sky Patrol Flying Saucer, batter-op toy, 7-1/2" diameter, KO Co., 1950s	150.00	225.00
Slide puzzle, alien with ship, 3-D slide puzzle, glows in dark, 4" x 4", DaMert Co., 1996	2.00	5.00
Space Ship X-5, tin and plastic battery-op toy, Japanese, late 1960s	150.00	250.00
Sticker book, UFOs–Past and Present, Whitman, 1968	20.00	35.00
Trading cards, "Mars Attacks," Bubbles-Topps, 1962		
card #1	30.00	40.00
card #2-54, each	15.00	20.00
card #55 (unmarked checklist)	75.00	100.00
full set of 55 cards	1,200.00	1,600.00

Push puppet, green alien, 5", Accoutrements, 1997-$4, MIP.

Plush toy, with plastic head, "cosmic sounds," eyes light up, 8-1/4", Gemmy Ind. Corp., 1996-$10.

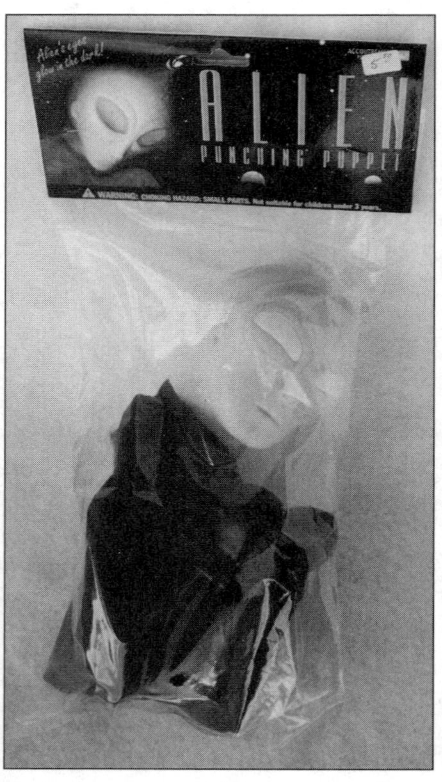

Push puppet, green alien, 5", Accoutrements, 1997-$4.

Puzzle, Visitors, shows alien head over pyramid, 1,000 pieces, 1990s-$22.

Item	Good/Loose	Mint/MIP
Trading cards, Mars Attacks Unpublished Version, ltd. ed. set, 13 cards, Rosem, 1984	35.00	45.00
Trading cards, Saucer People, boxed set of 36 cards, Kitchen Sink, 1992	10.00	15.00
Trading cards, UFO Trading Cards, boxed set of 50 cards, Dark Horse, 1998	12.00	16.00
UFO, battery-op spaceship, 5-1/4", Accoutrements, 1994	8.00	12.00
UFO, Beam Craft, 3-1/4" diameter, Shadow Box, 1997	6.00	10.00
UFO, Daylight Disk, 3-1/4" diameter, Shadow Box, 1997	6.00	10.00

Slide puzzle, alien with ship, 3-D slide puzzle, glows in dark, 4" x 4", DaMert Co., 1996-$5.

UFO, battery-op spaceship, 5-1/4", Accoutrements, 1994-$12.

Trading cards, Saucer People, boxed set of 36 cards, Kitchen Sink, 1992-$15.

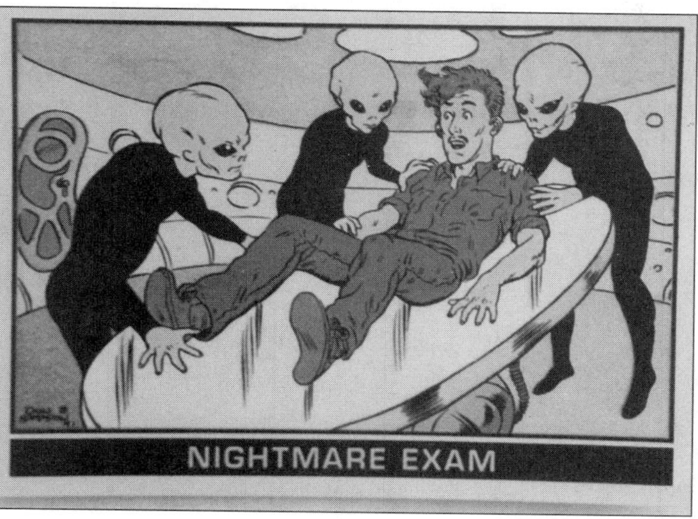

Item	Good/Loose	Mint/MIP
UFO, Mother Ship, 3-1/2", Shadow Box, 1997	6.00	10.00
UFO, Scout Craft, 3-1/4" diameter, Shadow Box, 1997	6.00	10.00
UFO, Triangle Craft, 3-1/2", Shadow Box, 1997	6.00	10.00
UFO Files alien bendee, Amphiboid Scientist, 5", Toy Concepts, 1997	5.00	8.00
UFO Files alien bendee, Arterian Captain, 5", Toy Concepts, 1997	5.00	8.00
UFO Files alien bendee, Galactic Commander, 5", Toy Concepts, 1997	5.00	8.00

UFO, Beam Craft, 3-1/4" diameter, Shadow Box, 1997-$10.

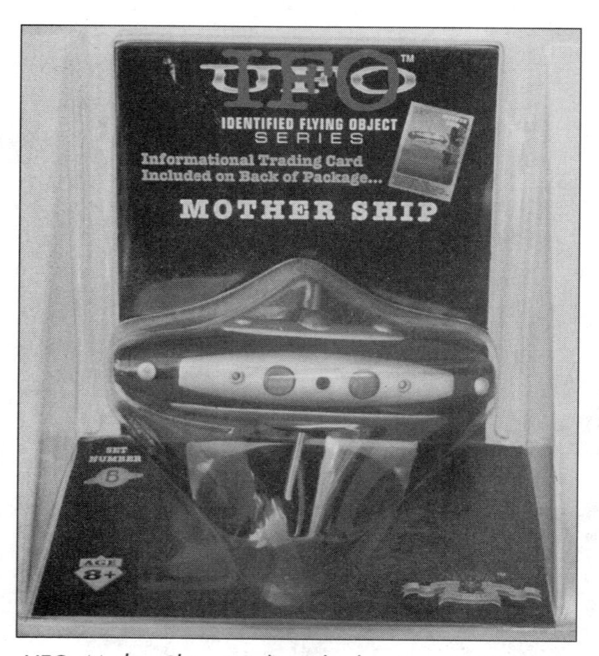

UFO, Mother Ship, 3-1/2", Shadow Box, 1997-$10.

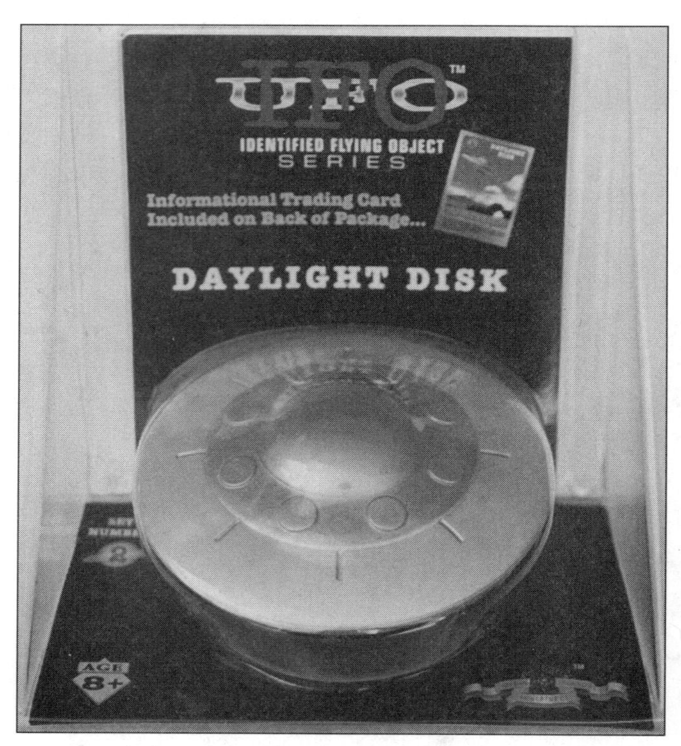

UFO, Daylight Disk, 3-1/4" diameter, Shadow Box, 1997-$10.

UFO, Mother Ship, 3-1/2", Shadow Box, 1997-$10.

Item	Good/Loose	Mint/MIP
UFO Files alien bendee, Grey Abductor, glow-in-the-dark eyes, 5", Toy Concepts, 1997	5.00	8.00
UFO Files alien bendee, Marshan Assassin, 5", Toy Concepts, 1997	5.00	8.00
UFO Files alien bendee, Muskel Eliminator, 5", Toy Concepts, 1997	5.00	8.00
UFO Mystery Ball, glows in dark, 2-1/2" across, OTC, 1990s, each	1.00	2.00
UFO XO5, battery op, moves, lights, beeps, Modern Toys, Masudaya of Japan, 1970s	60.00	120.00
Visitor (The), UFO with magnetic action, created by L.J. Teff & Assoc., DaMert Co., 1997	18.00	28.00
Waxy plastic figure, Milky Way, orange with dish head, 2", Miller, 1950s	125.00	175.00
Waxy plastic figure, Big Dipper, green, 2", Miller, 1950s	125.00	175.00
Waxy plastic figure, Mars, green with big head, 4", Miller, 1950s	125.00	175.00
Waxy plastic figure, Saturn, bird-like with gun, 4", Miller, 1950s	125.00	175.00
Waxy plastic figure, Pluto, dragon-like with blazing guns, 4" tall, Miller, 1950s	125.00	175.00
Waxy plastic figure, Purple People Eater, with silver trim, 4-3/4", Miller, 1950s	125.00	175.00
Waxy plastic figure, Venus, green painted plastic with red weapon, 4", Miller, 1950s	125.00	175.00
Wind-up walker, alien, glows in the dark, 2-1/4", Fame Masters Ltd., 1997	2.00	4.00
X-7 Space Explorer Ship, tin litho battery-op toy, Modern Toys, Japan, 1960s	250.00	350.00
X-15 Space Saucer, tin wind-up, 6" diameter, KO Co., Japan, 1950s	150.00	250.00
Yo-yo, Duncan Alien, glows in the dark, 3-1/2" diameter, 1990s	5.00	10.00
Yo-yo, metal, black with green alien, made in China, 1990s	1.00	2.00
Z-101 Flying Saucer, tin litho, Japanese, 1950s	75.00	125.00
Zem 21 space alien doll, 12", boxed, Ideal, 1970s	20.00	30.00

UFO, Triangle Craft, 3-1/2", Shadow Box, 1997-$10.

UFO Mystery Ball, glows in dark, 2-1/2" across, OTC, 1990s, each-$2.

UFO Files alien bendee, Grey Abductor, glow-in-the-dark eyes, 5", Toy Concepts, 1997-$8.

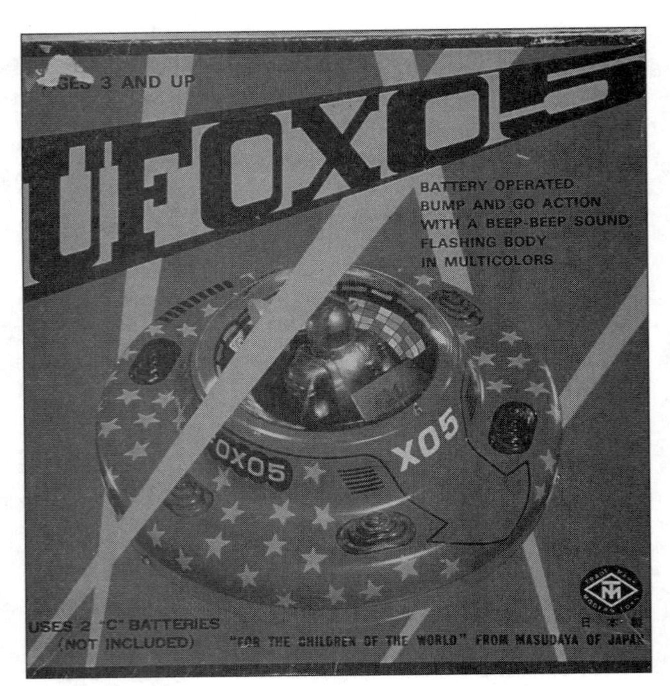

UFO XO5, battery op, moves, lights, beeps, Modern Toys, Masudaya of Japan, 1970s-$120.

Visitor (The), UFO with magnetic action, created by L.J. Teff & Assoc., DaMert Co., 1997-$28.

Wind-up walker, alien, glows in the dark, 2-1/4", Fame Masters Ltd., 1997-$4.

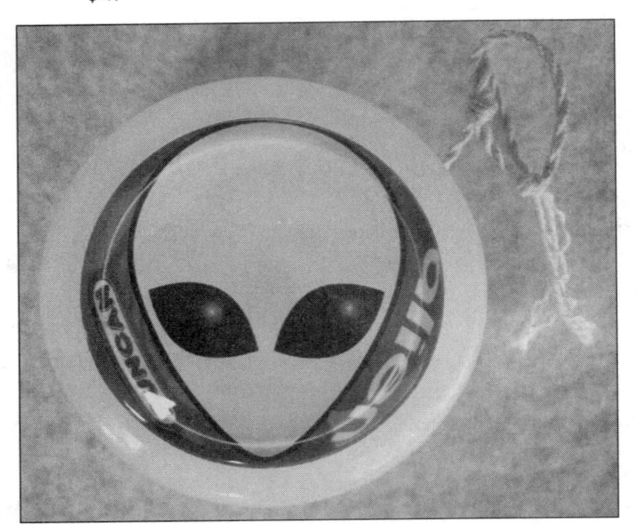

Yo-yo, Duncan Alien, glows in the dark, 3-1/2" diameter, 1990s-$10.

Yo-yo, metal, black with green alien, made in China, 1990s-$2.

Special Focus:

The Pascagoula UFO An Interview with Charles Hickson

Conducted by the author in November 1976
New Orleans, LA
Reprinted from "The UNO Driftwood"

"I'm not here to try to convince you that there's other worlds up there with life on it. I'm here, simply, to tell you a true story of what happened to Calvin and me on October 11, 1973 in Pascagoula, Mississippi."

So begins the strange story of Charles Hickson's visit inside a UFO. The center of one of the most famous and controversial UFO stories, Hickson was the special guest speaker at a UFO conference held downtown recently at the Foundation Faith of the Millennium.

Charles Hickson Jr. was there, too. Surprisingly, he had not heard his father tell the strange story before that night. "I was sent overseas ten days before it happened," he said, "...or I would have been fishing with them. I didn't find out about it 'til I picked up a paper in China one day and happened to read about it."

Since Hickson's son had been back home, he had not asked his father any questions about his mysterious abduction, nor had his father volunteered the information. So he decided to come hear his father speak in New Orleans to find out what it was really all about. He sat in the audience, intently interested as the rest of us were, and listened as Hickson described his adventure...

After leaving work late that afternoon, Charles Hickson and Calvin Parker went fishing on the Pascagoula River. "We tried two or three different places on the river, but the fish weren't biting, so we moved up river to a place where the fish always bite, by the Shaw Peter Shipyards. It was just about dark outside, and we went and sat down on the old pier there.

"I'm not sure what had attracted my attention," he continued, "whether I'd turned to get bait for my spinning reel or what, but I heard some kind of hissin' sound. It wasn't very loud, but it was quite shrill."

That's when Hickson looked up and saw the craft. It had blue flashing lights toward the front and top of it, and was hovering two or three feet above the ground. "I was startled," he told us. "I don't recall saying anything, but I remember by this time Calvin was standing up and looking at it too. Then an opening appeared in the craft. It didn't open like a door; it simply appeared. A light shined real brilliant out of the craft then, and the two blue lights on the outside simply went out."

Hickson said that at this point, he wasn't frightened, only curious. He didn't become really scared until he saw the "aliens." "These three creatures came out of the craft. They never touched the ground, but they hovered over it at about the same height as the craft did. It terrified me then. They just glided across the ground and they were instantly upon us."

Two of the creatures took Hickson's arms, and he recalls feeling pain in one arm for just a moment, perhaps like that from the prick of a needle. After that, he had "no feelings at all." One of the aliens got Cal as well, and Hickson remembers seeing his younger friend faint and grow limp.

"Then they carried us both inside," Hickson remembers. "We just sort of flowed inside of it. Then they picked us up off the ground to the same level that the creatures were."

They were taken into a round room and Hickson said there were no light fixtures. Instead, blindingly bright light emanated from the walls, ceiling and floor. Today, Hickson still has troubles with his eyes, believed to be due to these lights.

"They took me to the center of the room," Hickson went on, "...And something just seemed to be coming out of the wall. It was about the size of a football, and it stopped eight inches from my face. I couldn't move at all. I seemed to be frozen. I now refer to this thing as being like a big eye. It started to move down the front of me, like it was scanning me, and then it went under me. The next time I saw it, it was coming back up over my head, and when it got back in front of my face, it stopped."

Hickson says he could move only his eyes during this "examination." "I tried to wiggle my toes," he said, "but I couldn't. And they weren't holding me at this time. I was being suspended in the air somehow."

Hickson says the only sound he remembers was a slight buzzing sound, possibly coming from one of the creatures. But he does not try to offer an explanation for the sound.

Hickson said the fear he felt on board the alien ship was not "normal fear," defining normal fear as being something we're familiar with. "I was terrified," he said. "It almost frightened me to death."

When it was all over with, the creatures took the Pascagoulans back outside, and Hickson remembers being released from their hold and falling to the ground.

"I looked up and saw Calvin standing, facing the river with his arms outstretched," he said. "And I wondered if he was in shock."

Hickson crawled over to his friend to try to help him and remembers it was quite a while before he could get to talk to him or make any sense out of what he was saying.

When Cal was back to normal, the question came up: What were they going to do? At first, they decided to try to forget about the whole thing. "We didn't want to be called 'nuts,'" Hickson said. But they eventually realized that some authorities should be told their story. They contacted the Jackson County sheriff, convinced him they weren't lying and told him not to leak anything to the press. The next day at the shipyard, calls for Charles Hickson jammed the phone lines. The news had leaked to the press.

Now that everyone knows the story, Hickson sees no reason to try to hide it. He is publishing a book sometime next month called *The Pascagoula UFO*. This book will include artists' sketches of the aliens and their ship, recreated from Hickson's subconscious memory and revealed through a series of hypnotic regression sessions.

Hickson now believes that these "creatures" were actually robots. He says there was no movement at all in the chest area or the mouth area to imply breathing. Also, they moved very mechanically, their shoulders and leg joints being the only moving parts.

One of the things that Hickson mentioned, but declined to discuss, was the fact that he believes the aliens are still watching him. "I feel like they know everything I'm doing," he says. "But, I'm convinced that they mean me no harm. I'm no longer afraid of them."

In conclusion, Hickson pointed out that he is a firm believer in God, a hard-shelled Baptist. "I was raised to believe that God created the universe," he says. "I still believe that God is the master of other worlds, as sure as he's the master of this world."

Chapter 2
Books, Periodicals and Other Sacred Texts

One of the most active and interesting areas of UFO and alien collecting is literature. UFO buffs and collectors are typically hungry for information on what is going on, what others think is going on and how various groups are attempting to explain, prove or debunk it all.

Many UFO books can be separated into the following basic categories: Entity Encounters, Parapsychic Explanations, Regional Studies, Nuts and Bolts saucer books, Government Cover-Ups and Debunkers. Entity Encounter books usually describe contact with an alien, either through physical or mental channels, without necessarily trying to explain how or why it happened—they just document the action and any information that may have been transferred. Parapsychic Explanations, particularly popular in Europe today, tend to favor the notion that these may not be ships and people from space; it's probably something really strange from closer by. Regional Studies tend to be fairly objective investigations into various sightings and incidents, often by an outside party seeking an explanation and carefully documenting the events. The Nuts and Bolts saucer books are typically written by scientists trying to make sense of the phenomenon as a whole, looking at various cases and attempting to decipher the alien intent, origin and transportation technology. Government cover-up books and most of the debunker books basically fire the ongoing debate "Is it real or mass delusion?"

Early editions of works by some of the field's pioneers are especially collectible. In the 1950s, retired Air Force major Donald E. Keyhoe became one of the most respected writers on flying saucers. At about the same time, George Adamski stunned an often-disbelieving audience with his tales of travels to Venus and meeting with Venusians on board their spacecraft. In the 1960s, John Fuller wrote the first publicized account of an alien abduction, *The Interrupted Journey*, about Betty and Barney Hill. Other major authors in the field include Erik Von Daniken (*Chariots of the Gods*), Jacques Vallee, John Keel, J. Allen Hynek, Phillip Klass (the king debunker), and more recently Whitley Streiber (*Communion*) and Ed Walters (*The Gulf Breeze Sightings*).

UFO and alien-themed periodicals and magazines have also developed a following of fans and collectors. The earliest publications, including *Fate*, which debuted in 1948 with a flying saucer cover story, are highly sought after and quite valuable. Most magazines and newsletters from the 1950s and '60s, if found in good shape, fetch $10-$20 or more today. Pre-1990 issues of British import magazines are also very desirable to collectors in the U.S. Some titles, such as *True*, *Saga* and *Argosy*-related UFO magazines, may be boosted in value by the cross-over "men's magazine" collector market.

Flying Saucer Review (FSR), published in England since 1955, is one of the oldest existing periodicals. It has taken the lead in portraying UFOs as a parapsychic, new age-style phenomenon. Another early publication, *Saucer News*, debuted in 1954 under the title *Nexus*. Edited by James W. Mosely, it was irreverent, controversial and often light-hearted in tone. A personal 'zine, rather than a slick publication, *Saucer News* and its more recent incarnation (since 1981), *Saucer Smear*, has typically been traded, not sold, so it is valuable to collectors. Mosely also takes particular delight in gossiping about other UFO researchers—a fun read.

Most UFO magazines have their own distinctive flavor. One title, *Official UFO*, published in the mid-1970s, began as a very respectable periodical, with big-name authors and lots of credibility. Then, around 1977, it suddenly began printing "suspicious material"; the stories became more and more outlandish in later issues. Finally, its downfall was sealed when the April 1978 cover story claimed that a UFO had destroyed an entire Illinois city and reconstructed it overnight. When skeptics asked citizens what they'd seen, the outrageous hoax was exposed and the magazine folded shortly thereafter.

One of the most active editors in the UFOlogy periodical field today is Timothy Green Buckley, whose titles include *Alien Encounters, UFO Universe, Uncensored UFO Reports, Unexplained Universe* and *Unsolved UFO Sightings*.

Books

Item	Good	Mint
Abducted!, Debbie Jordan and Kathy Mitchell, Dell Books, 1995	3.00	5.00
Abduction, John Mack, hardcover, Macmillan, 1994	18.00	22.00
Abductions: A Dangerous Game, Phillip Klass, hardcover, Prometheus Books, 1988	16.00	22.00

Alien Identities, Richard Thompson, trade paperback, Govardhan Hill Pub., 1993-$15.

Aliens From Space..., Major Donald E. Keyhoe, hardcover, Doubleday Book Club Ed., 1973-$12.

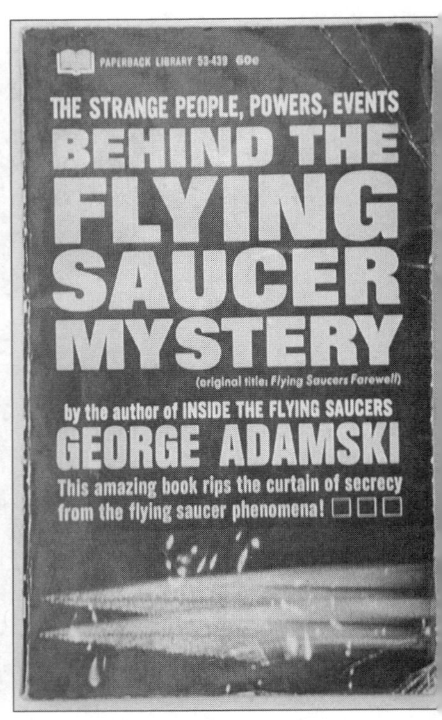

Behind the Flying Saucer Mystery, George Adamski, Paperback Library, 1967-$15.

Item	Good	Mint
Above Top Secret, Timothy Good, 1st Ed. hardcover, William Morrow & Co., 1988	15.00	25.00
Advanced Aerial Devices Reported During the Korean War, Richard Haines, softcover, LDA Press, 1990	8.00	12.00
Alien Abductions—Mystery Solved, Jenny Randles, trade paperback, Inner Light, 1988	10.00	15.00
Alien Abduction, UFOs & the Conference at MIT, C.D.B. Bryan, hardcover, Borzoi Books	20.00	25.00
Alien Contact, Timothy Good, 1st U.S. edition, hardcover, Morrow, 1993	20.00	25.00
Alien Identities, Richard Thompson, trade paperback, Govardhan Hill Pub., 1993	10.00	15.00
Alien Jigsaw, The, Katherine Wilson, hardcover, Puzzle Publishing, 1993	18.00	28.00
Alien Tide, The, Tom Dongo, trade paperback, Hummingbird Press, 1990	4.00	8.00
Aliens Above, Always, John Magor, trade paperback, Hancock House, 1983	8.00	12.00
Aliens: Encounters with the Unexplained, Marcus Day, hardcover, CLB, 1997	8.00	15.00
Aliens From Space..., Major Donald E. Keyhoe, hardcover, Doubleday Book Club Ed., 1973	8.00	12.00
Aliens From Space..., Major Donald E. Keyhoe, softcover, Signet Books, 1973	3.00	5.00
Aliens in the Skies, John Fuller, hardcover, G.P. Putnam's Sons, 1969	12.00	20.00
An Account of Meeting with Denizens of Another World 1871, William Robert Loosley, hardcover, St. Martin's Press, 1979/80	6.00	10.00

Behind the Flying Saucers, Frank Scully, Popular Library paperback, 1951-$15.

Item	Good	Mint
Anatomy of a Phenomenon, Jacques Vallee, hardcover, Henry Regnery Co., 1965	15.00	25.00
Andreasson Affair: Phase II, The, Raymond Fowler, hardcover, Prentice-Hall, 1982	8.00	15.00
Angels and Aliens, Keith Thompson, hardcover, Addison Wesley, 1991	15.00	20.00
Are the Invaders Coming?, Steven Tyler, softcover, Tower Books, 1968	5.00	8.00
Aurora, the Pentagon's Secret Hypersonic Spyplane, Bill Sweetman, softcover, 1993	4.00	8.00

Close Extraterrestrial Encounters, R. and L. Boylan, trade paperback, Wildflower Press, 1994-$15.

Cosmic Pulse of Life, The, T.J. Constable, trade paperback, Merlin Press, 1976-$40.

Beyond My Wildest Dreams, Kim Carlsberg, illustrated trade paperback, Bear & Co., 1995-$20.

Item	Good	Mint
Behind the Flying Saucer Mystery, George Adamski, Paperback Library, 1967	10.00	15.00
Behind the Flying Saucers, Frank Scully, Popular Library paperback, 1951	8.00	15.00
Beyond Earth: Man's Contact with UFOs, Ralph and Judy Blum, Bantam paperback, 1974	3.00	5.00
Beyond My Wildest Dreams, Kim Carlsberg, illustrated trade paperback, Bear & Co., 1995	15.00	20.00
Cash Landrum UFO Incident, The, John Schuessler, Geo Graphics Printing, 1998	15.00	20.00
Chariots of the Gods?, Erich Von Daniken, U.S. Bantam paperback, 1970s	1.00	3.00
Clear Intent (Govt. Cover-up of UFOs), Fawcett and Greenwood, softcover, Prentice Hall, 1984	5.00	8.00
Close Extraterrestrial Encounters, R. and L. Boylan, trade paperback, Wildflower Press, 1994	10.00	15.00
Communion: A True Story, Whitley Strieber, 1st Ed. hardcover, BeechTree Books, 1987	20.00	25.00
Communion: A True Story, Whitley Strieber, Avon paperback, 1987	2.00	5.00
Complete Book of UFOs, The, Jenny Randles and Peter Hough, softcover, Sterling, 1996	10.00	18.00
Confirmation, Whitley Strieber, hardcover, St. Martin's Press, 1998	18.00	24.00
Confrontations, Jacques Vallee, hardcover, Ballantine, 1990	10.00	20.00
Contact of the 5th Kind, Philip J. Imbrogno, Llewellyn Publishing, 1997	7.00	10.00
Cosmic Pulse of Life, The, T.J. Constable, trade paperback, Merlin Press, 1976	30.00	40.00
Cosmic Voyage, Courtney Brown, 1st Ed. hardcover, Dutton Books, 1996	18.00	15.00
Crash at Corona, Friedman and Berliner, hardcover, Paragon House, 1992	15.00	25.00
Dimensions, Jacques Vallee, hardcover, Contemporary Books, 1988	15.00	20.00
Disneyland of the Gods, John Keel, rade paperback, Amok Press, 1988	7.00	15.00

Extra-Terrestrials on Earth, Joshua Strickland, trade paperback, Grosset & Dunlap, 1977-$12.

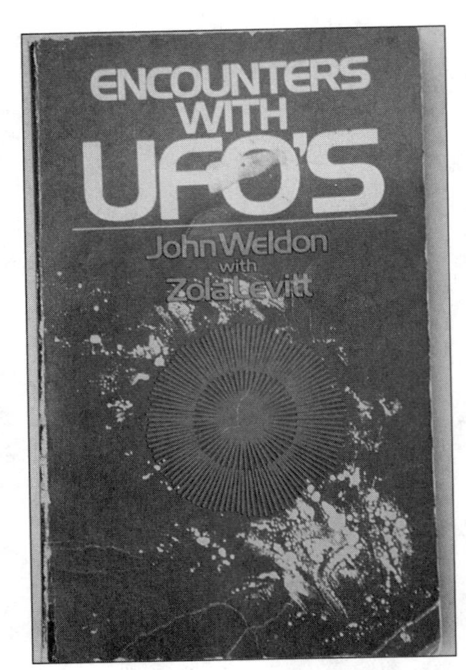

Encounters With UFOs, Weldon and Levitt, softcover, Harvest House, 1978-$8.

Extra-Terrestrials Among Us, George Andrews, trade paperback, Llewellyn Pubs., 1987-$15.

Fire in the Sky, Travis Walton, (film tie-in), hardcover, Marlowe & Co., 1996-$25.

Item	Good	Mint
Dragon and the Disc, The, E.W. Holiday, Norton, 1993	15.00	25.00
Dyfed Enigma, The, Randle Jones Pugh, Faber and Faber Press, 1979	10.00	15.00
Earth Lights Revelations, Paul Devereux, trade paperback, Blandford Press, 1989	8.00	15.00
Edge of Reality, The, J. Allen Hynek and Jacques Vallee, softcover, Henry Regnery Co., 1975	12.00	20.00
Encounters, Edith Fiore, hardcover, Doubleday, 1989	15.00	20.00
Encounters With UFOs, Weldon and Levitt, softcover, Harvest House, 1978	5.00	8.00
Erinnerungen an die Zikunft (original German, *Chariots of the Gods*), Erich Von Daniken, Econ-Verlag, 1968	30.00	50.00

Item	Good	Mint
Examining the Earthlight Theory, Greg Long, softcover, Priority Publishing, 1990	5.00	8.00
Extra-Terrestrial Friends and Foes, George Andrews, trade paperback, IllumiNet Press, 1993	14.00	18.00
Extra-Terrestrials Among Us, George Andrews, trade paperback, Llewellyn Pubs., 1987	10.00	15.00
Extra-Terrestrials on Earth, Joshua Strickland, trade paperback, Grosset & Dunlap, 1977	6.00	12.00
Faces of the Visitors, Kevin Randle and Russ Estes, softcover, Simon & Schuster, 1997	10.00	12.00
Field Guide to Extraterrestrials, The, Patric Huyghe, trade paperback, Avon, 1996	10.00	13.00
Fire in the Sky, Travis Walton, (film tie-in), hardcover, Marlowe & Co., 1996	20.00	25.00

Flying Saucers, Dr. C.G. Jung, Signet paperback, 1969-$12.

Flying Saucers and the Three Men, Albert Bender, Paperback Library, 1968-$8.

Flying Saucers on the Attack, Harold Wilkins, softcover, Ace Star Book, 1967-$12.

Item	Good	Mint
Flying Saucer Conspiracy, The, Donald E. Keyhoe, 1st Ed., Henry Holt & Co., 1955	15.00	25.00
Flying Saucers, Dr. C.G. Jung, Signet paperback, 1969	8.00	12.00
Flying Saucers and the Straight-Line Mystery, Aime Michel, hardcover, S.G. Phillips, 1958	15.00	20.00
Flying Saucers and the Three Men, Albert Bender, Paperback Library, 1968	5.00	8.00
Flying Saucers Are Hostile, Steiger and Whritenour, softcover, Award Books, 1967	5.00	8.00
Flying Saucers Are Real, The, Donald Keyhoe, 1st printing, Fawcett Gold Medal, 1950	20.00	30.00
Flying Saucers From Outer Space, Donald Keyhoe, hardcover, Henry Holt & Co., 1953	20.00	25.00
Flying Saucers on the Attack, Harold Wilkins, softcover, Ace Star Book, 1967	8.00	12.00
Flying Saucers: The Startling Evidence of the Invasion from Outer Space, Coral Lorenzen, softcover, Signet, 1966	8.00	12.00
Flying Saucers: Top Secret, Donald Keyhoe, hardcover, G.P. Putnam & Sons, 1960	25.00	45.00
Flying Saucers Uncensored, Harold Wilkins, hardcover, Citadel Press, 1955	15.00	25.00
Forbidden Science (Journals 1957-69), Jacques Vallee, hardcover, North Atlantic Books, 1992	20.00	30.00
From Out of the Blue, Jenny Randles, Berkley paperback, 1993	3.00	5.00
Gods from Outer Space, Erich Von Daniken, Bantam paperback, 1972–73	1.00	3.00

Letters to the Air Force on UFOs, Bill Adler, softcover, Dell Books, 1967-$9.

Item	Good	Mint
Gold of the Gods, The, Erich Von Daniken, Bantam paperback, 1974	1.00	3.00
Great Flying Saucer Hoax, Carol Lorenzen, 1962	12.00	20.00
Gulf Breeze Sightings, The, Ed and Frances Walters, Morrow hardcover, 1990	20.00	25.00
Hands, (channeling alien entity), Margaret Williams and Lee Gladen, Galaxy Press, 1970s	5.00	10.00
Helicopter-UFO Encounter Over Ohio (A), Jennie Zeidman, 8.5 x 11 in. report format, published by Center for UFO Studies	12.00	15.00
House of Lords UFO Debate, The, Lord Clancarty, softcover, Open Head Press, 1979	12.00	20.00
Incident at Exeter, John Fuller, G.P Putnam's Sons	10.00	25.00
Inside Saucer Post... 3-0 Blue, L. Stringfield, softcover, published by Civilian Research, *Interplanetary Flying Objects*, 1st Ed., October 1957	12.00	20.00
Inside the Flying Saucers, George Adamski, Warner Paperback Library, 1967	10.00	15.00
Inside the Space Ships, George Adamski, hardcover, Abelard-Schwan Ink, 1955	35.00	50.00
Interrupted Journey, The, (about Betty & Barney Hill), John Fuller, Dile Press, 1965	10.00	25.00

Is Another World Watching?, Gerald Heard, Bantam paperback, 1953-$8.

Inside the Flying Saucers, George Adamski, Warner Paperback Library, 1967-$15.

Flying Saucers Are Hostile, Steiger and Whritenour, soft-cover, Award Books, 1967-$8.

Gulf Breeze Sightings, The, Ed and Frances Walters, Morrow hard-cover, 1990-$25.

Item	Good	Mint
Intruders, Budd Hopkins, 1st Ed. hardcover, Random House, 1987	15.00	20.00
Invisible College, The, Jacques Vallee, 1st Ed. hardcover, E.P. Dutton & Co., 1975	4.00	8.00
Invisible Residents, Ivan Sanderson, hardcover, World Publishing Co., 1970	8.00	15.00
Is Another World Watching?, Gerald Heard, Bantam paperback, 1953	5.00	8.00
Is something up there?, Dale White, Scholastic, 1968	3.00	5.00
Left at East Gate, Larry Warren and Peter Robbins, hardcover, Marlow & Co., 1997	15.00	25.00
Let's Face the Facts About Flying Saucers, Gabriel Green, softcover, Popular Library, 1967	5.00	8.00
Letters to the Air Force on UFOs, Bill Adler, softcover, Dell Books, 1967	5.00	9.00
Making Contact, Bill Fawcett, hardcover, Morrow Press, 1997	18.00	23.00
Melbourne Episode, Richard Haines, softcover, L.D.A. Press, 1987	8.00	15.00
Missing Time, Budd Hopkins, hardcover, Richard Marek Publishers, 1981	12.00	20.00
Mothman Prophecies, The, John Keel, Saturday Review Press, 1975	20.00	30.00
Mysteries of Sedona, The, Tom Dongo, trade paperback, Hummingbird Press, 1988	4.00	8.00
Mysterious Valley, The, Christopher O'Brien, St. Martin's Paperbacks, 1996	4.00	7.00
National Enquirer UFO Report, Pocket Books paperback, 1985	2.00	5.00
Night Siege, Imbrogno, Pratt and Hynek, trade paperback, Llewellyn Publications, 1998	8.00	10.00
Night Siege—The Northern Ohio UFO-Creature Invasion, Dennis Pilichis, zine format, 1982	10.00	15.00
Not of This World, Peter Kolosimo, Bantam paperback, U.S. version, 1970s	3.00	5.00

Let's Face the Facts About Flying Saucers, Gabriel Green, softcover, Popular Library, 1967-$8.

Mysterious Valley, The, Christopher O'Brien, St. Martin's Paperbacks, 1996-$7.

National Enquirer UFO Report, Pocket Books paperback, 1985-$5.

Observing UFOs, Richard Haines, trade paperback, Nelson-Hall Publishers, 1980-$12.

Out There, Howard Blum, hardcover, Simon & Schuster, 1990-$25.

Passport to Magonia, Jacques Vallee, trade paperback, Henry Regnery Co., 1969-$20.

Project Blue Book, edited by Brad Steiger, Ballantine paperback, 1987-$5.

Item	Good	Mint
Observing UFOs, Richard Haines, trade paperback, Nelson-Hall Publishers, 1980	6.00	12.00
Official Guide to UFOs, The, *Science & Mechanics* editors, Ace Books paperback, 1968	3.00	8.00
On Pilots and UFOs, Willy Smith, trade paperback, Unicat Project, 1997	10.00	13.00
Open Skies, Closed Minds, Nick Pope, hardcover, Simon & Schuster, 1996	15.00	25.00
Out There, Howard Blum, hardcover, Simon & Schuster, 1990	15.00	25.00
Passport to Magonia, Jacques Vallee, trade paperback, Henry Regnery Co., 1969	12.00	20.00
Philadelphia Experiment & Other UFO Conspiracies, The, Brad Steiger, softcover, Inner Light Publications, 1990	6.00	10.00
Project Blue Book, edited by Brad Steiger, Ballantine, 1st. Ed., 1976	7.00	12.00
Project Blue Book, edited by Brad Steiger, Ballantine paperback, 1987	2.00	5.00
Project Delta: A Stud of Multiple UFOs, Richard Haines, LDA Press, 1994	10.00	20.00
Project Identification, Harley Rutledge, hardcover, Prentice Hall, 1981	10.00	20.00
Quest, The, Tom Dongo, trade paperback, Hummingbird Press,1990s	5.00	9.00
Report on Communion, Ed Conroy, hardcover, Morrow, 1989	15.00	20.00
Report on Unidentified Flying Objects, The, E. Ruppelt, Doubleday Book Club Ed., 1956	10.00	25.00
Return to the Stars, Erich Von Daniken (original English title), Souvenir Press, 1970	7.00	12.00

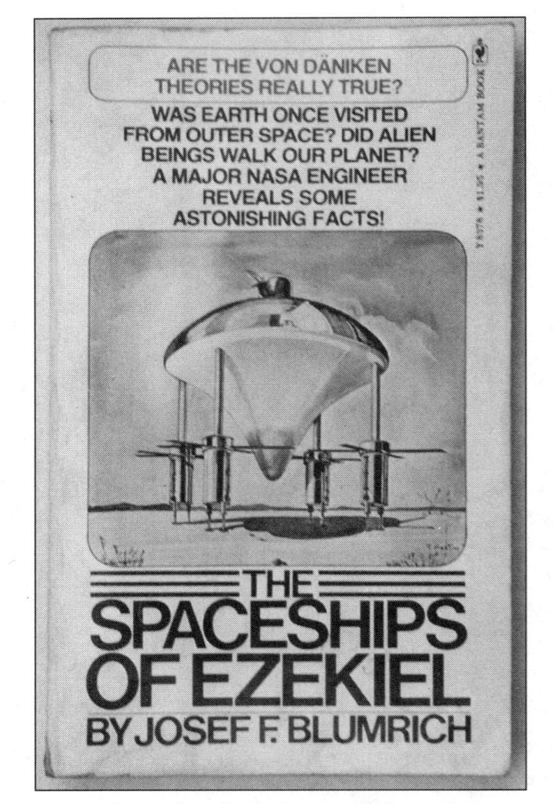

Spaceships of Ezekiel, The, Josef Blumrich, Bantam paperback, 1974-$7.

Strange Creatures from Time and Space, John Keel, softcover, Fawcett, 1970-$15.

Strangers from the Skies, Brad Steiger, softcover, Award Books, 1966-$10.

Truth About Flying Saucers, The, Aime Michel, softcover, Pyramid Books, 1974-$3.

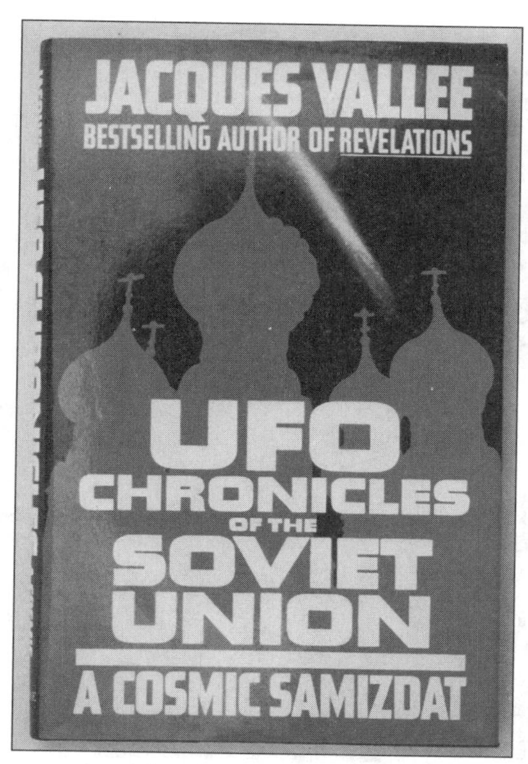

UFO Chronicles of the Soviet Union, Jacques Vallee, 1st Ed. hardcover, Ballantine, 1992-$20.

Item	Good	Mint
Revelations, Jacques Vallee, hardcover, Ballantine, 1991	15.00	20.00
Sky Creatures: Living UFOs, Trevor J. Constable, softcover, Pocket Books, 1978	3.00	5.00
Sky People, The, Brinsly Le Poer Trench, British softcover, Tandem, 1971	10.00	15.00
Spaceship Returns, The, Beman Lord, children's book, Henry Z. Walck, 1970	2.00	5.00
Spaceships of Ezekiel, The, Josef Blumrich, Bantam paperback, 1974	4.00	7.00
Strange Creatures from Time and Space, John Keel, softcover, Fawcett, 1970	10.00	15.00
Strangers from the Skies, Brad Steiger, softcover, Award Books, 1966	5.00	10.00
Timeless Earth, Peter Kolosimo, Bantam paperback, U.S. version, 1975	3.00	5.00
Truth About Flying Saucers, The, Aime Michel, hardcover, Criterion Books, 1956	12.00	20.00
Truth About Flying Saucers, The, Aime Michel, softcover, Pyramid Books, 1974	1.00	3.00
Tuhinga Canyon Contacts, The, Ann Druffel and Scott Rogo, hardcover, Prentice-Hall, 1980	4.00	10.00
UFO, Robert Chapman, softcover, Mayflower Books (British), 1974	6.00	10.00
UFO Abductions in Gulf Breeze, Ed and Francis Walters, softcover, Avon, 1994	3.00	6.00
UFO Chronicles of the Soviet Union, Jacques Vallee, 1st Ed. hardcover, Ballantine, 1992	15.00	20.00

Item	Good	Mint
UFO Contact from Undersea, Sanchz-Ocejo and Stevens, ltd. 1st Ed. of 3,000, UFO Photo Archives, 1982	15.00	25.00
UFO Controversy in America, The, D.M. Jacobs, hardcover, Indiana University Press, 1975	12.00	20.00
UFO Crash at Aztec, Steinman and Stevens,ltd. 1st edition, 1,000 copies, hardcover, 1986	30.00	50.00
UFO Evidence, The, edited by Richard Hall, published by NICAP, softcover, 1964	25.00	35.00
UFO Experience, The: A Scientific Inquiry, J. Allen Hynek, hardcover, Henry Regnery Co., 1972	15.00	20.00
UFO—Ho Ho!, cartoons by Joseph Farris, Popular Library, 1968	1.00	4.00
UFO Occupants & Critters, John B. Musgrave, zine format, Global Communications, 1979	6.00	9.00
UFO Photographs, Vol. 1, Stevens and Roberts, hardcover, UFO Photo Archives, 1986	15.00	25.00
UFO Photographs, Vol. 2, Stevens and Roberts, hardcover, UFO Photo Archives, 1985	15.00	25.00
UFO Question (Not Yet Answered), The, P.J. Willcox, hardcover, Libra Publishers, 1976	7.00	10.00
U.F.O. Report, The, Irving Greenfield, softcover, Lancer Books, 1967	5.00	8.00
UFO Sightings, Landing and Abductions, Yurko Bondarchuk, softcover, Methuen, 1979	12.00	20.00
UFO Silencers, The, (about MIB), Timothy G. Beckley, trade paperback, Inner Light, 1990	8.00	12.00
UFOlogy, James McCampbell, trade paperback, Celestial Arts, 1976	8.00	12.00
UFOs 1947–1987, Hilary Evans, trade paperback, Fortean Tomes (British), 1987	20.00	30.00

UFOlogy, James McCampbell, trade paperback, Celestial Arts, 1976-$12.

UFO Crash at Aztec, Steinman and Stevens, ltd. 1st edition, 1,000 copies, hardcover, 1986-$50.

UFO—Ho Ho!, cartoons by Joseph Farris, Popular Library, 1968-$4.

U.F.O. Report, The, Irving Greenfield, softcover, Lancer Books, 1967-$8.

Item	Good	Mint
UFOs: A Scientific Debate, Carl Sagan and T. Page, softcover, Norton Library, 1974	15.00	25.00
UFOs and Their Mission Impossible, Dr. Clifford Wilson, Signet paperback, 1974	2.00	4.00
UFOs are Real: Here's the Proof, Ed Walters and Bruce Maccabee, softcover, Avon, 1997	3.00	6.00
UFOs Over the Americas, Jim and Coral Lorenzen, Signet Paperback, 1967	5.00	8.00
UFOs: The Whole Story, Coral and Jim Lorenzen, Signet paperback, 1969	5.00	8.00
UFOs Trojan Horse, John Keel, hardcover, Putnam, 1970	30.00	50.00
UFOs? Yes!, D. Saunders and R. Harkins, hardcover, 1st Ed., World Publishing Co., 1969	12.00	20.00
Ultimate Encounter, (about Travis Walton), Bill Barry, softcover, Pocket Books, 1978	1.00	3.00
Unconventional Flying Objects, Paul Hill, softcover, Hampton Roads Publishing, 1995	8.00	16.00
Uninvited, The, Nick Pope, Simon and Schuster, 1997	12.00	16.00
Uninvited Visitors, Ivan Sanderson, hardcover with holographic cover, Cowles, 1967	25.00	40.00
Utah UFO Display, The, Frank Salisbury, hardcover, Devin-Adair Comp., 1974	17.00	30.00
Walton Experience, The, Travis Walton, softcover, Berkley Medallion, 1978	2.00	5.00
West Virginia UFOs, Bob Teets, trade paperback, Headline Books, 1995	15.00	20.00
Witnessed, Budd Hopkins, 1st Ed. hardcover, Pocket Books, 1996	20.00	25.00
World's Best "True" UFO Stories, Randles ad Hough, hardcover, Sterling Pub. Co., 1994	8.00	12.00
Zip-Zip and His flying Saucer, John Schealer, children's book, 1st Ed., Dutton, 1956	3.00	6.00

UFOs 1947-1987, Hilary Evans, trade paperback, Fortean Tomes (British), 1987-$30.

UFOs and Their Mission Impossible, Dr. Clifford Wilson, Signet paperback, 1974-$4.

UFOs Over the Americas, Jim and Coral Lorenzen, Signet Paperback, 1967-$8.

Uninvited Visitors, Ivan Sanderson, hardcover with holographic cover, Cowles, 1967-$40.

Magazines, Newsletters and other Periodicals

Item	Good	Mint
Alien Encounters (UFO's Alien Encounters), published by GCR Publishing, New York; Editor, Timothy Green Beckley, 1970s-present		
#1	6.00	12.00
Most 1970s issues	5.00	10.00
Most 1980s issues	4.00	8.00
Most 1990s issues	3.00	5.00
#2, Collector's Edition, "Govt. Agents Collaborate with Aliens to Enslave Earth," 1995	3.00	5.00
A.P.R.O. Bulletin, The, published by the Aerial Phenomena Research Organization, Tucson; Editor, Coral E. Lorenzen, 1952-1988		
#1 (newsletter format)	15.00	20.00
Most 1950s issues	12.00	18.00
Most 1960s issues	10.00	15.00
Most 1970s issues	8.00	12.00
Most 1980s issues	5.00	8.00
Vol. 29, #10, September 1981	5.00	8.00
Argosy, "Flying Saucers Invade Finland" cover story, October 1971	8.00	12.00
Argosy UFO (see *UFO*)		

the apro bulletin

NEW MEXICO REPEAT REPORTS

UFO SCOFFER OBERG AND HIS "YARDSTICK"

A.P.R.O. Bulletin, The, Vol. 29, #10, September 1981-$8.

Beyond (From Beyond), Vol. 1, #4-$10.

Alien Encounters (UFO's Alien Encounters)-$5.

Argosy, "Flying Saucers Invade Finland" cover story, October 1971-$12.

Beyond, Vol. 2, #14, "Alien UFO Destroys Dog in Utah," October 1969-$20.

Beyond (From Beyond), Vol. 1, #10-$10.

Item	Good	Mint
Beyond, published by Beyond, Inc., Hicksville, NY		
Editor, Keith Ayling, 1960s-1970s		
#1	15.00	25.00
Most 1960s issues	12.00	20.00
Most 1970s issues	10.00	15.00
Vol. 2, #14, "Alien UFO Destroys Dog in Utah," October 1969	12.00	20.00
Beyond (From Beyond), published by Rapide Publishing, England		
Editor, Richard Forsythe, 1990s		
Most issues	6.00	10.00
Vol. 1, #4, "Cover-Up Revealed"	6.00	10.00
Vol. 1, #10, "Alien Embryo"	6.00	10.00
Beyond Reality, published by Beyond Reality Magazine, Inc., New York		
Editor/Publisher, Harry Belil, 1970s-1980		
Most issues	5.00	10.00
#28, "Special Issue! UFO Update!," Sept./Oct. 1977	6.00	9.00
Canadian UFO Report, published by Rapier Press Ltd.		
Editor/Publisher, John Magor, 1970s		
Most issues	10.00	25.00
Vol. 2, #4, "The Best UFO Photo Ever Taken," 1971	15.00	23.00
Excluded Middle, The		
Editor/Publisher, Gregory Bishop, 1990s		
#1	12.00	16.00
Most other issues	6.00	10.00
Far Out (California Far Out), published by Larry Flynt/LFP Inc., Beverly Hills, CA		
Editor, Michael DiGregorio, 1990s (discontinued)		
All issues	10.00	15.00
Vol. 2, #VI, "NASA's Search for Alien Life, Part 2," Winter 1993	10.00	15.00

Beyond Reality, #28, "Special Issue! UFO Update!," Sept./Oct. 1977-$9.

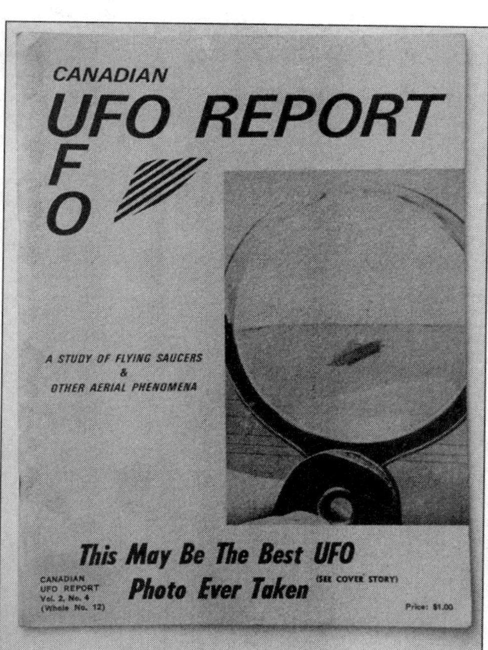

Canadian UFO Report, Vol. 2, #4, "The Best UFO Photo Ever Taken," 1971-$23.

Far Out (California Far Out), Vol. 2, #VI, "NASA's Search for Alien Life, Part 2," Winter 1993-$15.

Item	Good	Mint
Fate, published by Clark Publishing Co. 1948-1988, Llewellyn Publications Dec. 1988-present Editors, various over the years, Spring 1948-present		
Vol. 1, #1, "The Flying Disks," Spring 1948	40.00	70.00
Most 1940s issues	20.00	30.00
Most 1950s issues	10.00	18.00
Most 1960s issues	8.00	12.00
Most 1970s issues	6.00	10.00
Most 1980s issues	5.00	8.00
Most 1990s issues	3.00	6.00
Vol. 1, #2, "Crow River Flying Disk?" Summer 1948	20.00	30.00
Vol. 4, #1, "Easter Island," January 1951	15.00	20.00
Vol. 7, #5, "Hunt for the Saucers," May 1954	10.00	18.00
Vol. 10, #4, "I Have Seen Zombies," April 1957	10.00	18.00
Vol. 10, #6, "Mystery of the Green Fireballs," June 1957	10.00	18.00
Vol. 10, #8, "Saucers Over Europe," August 1957	10.00	18.00
Vol. 10, #10, "Italian Flying Saucer, October 1957	10.00	18.00
Vol. 11, #2, "Special Space Issue," February 1958	15.00	20.00
Vol. 11, #3, "Man is on the Road to the Stars," March 1958	10.00	18.00
Vol. 45, #3, "Men in Black," March 1992	2.00	4.00
Vol. 45, #9, "UFO Cover-Up," Sept. 1993	2.00	4.00

Fate, Vol. 4, #1-$20.

Fate, Vol. 7, #5-$18.

Fate, Vol. 10, #4-$18.

Fate, Vol. 10, #8-$18.

MAN IS ON THE ROAD TO THE STARS!
FATE
MAGAZINE
March 1958
WILLY LEY — Our Timetable To Space
MAX MILLER — Mars, The Mystery Planet
FRANK EDWARDS — Sightings and Alibis

Top L to R: Fate, Vol. 10, #10-$18. Fate, Vol. 11, #2-$20. Middle: Fate, Vol. 11, #3-$18. Bottom L to R: Fate, Vol. 45, #3-$4. Fate, Vol. 45, #9-$4.

Item	Good	Mint
Flying Saucer Menace, The, published by Universal Publishing/Universal-Tandem Ltd. By Brad Steiger and August Roberts, one-shot, 90 photographs, 1967	15.00	23.00
Flying Saucer Review (FSR), published by FSR Publications, England Editors, various, 1955-present		
#1, 1955	35.00	55.00
Most 1950s issues	25.00	35.00
Most 1960s issues	18.00	28.00
Most 1970s issues	15.00	20.00
Most 1980s issues	12.00	18.00
Most 1990s issues	10.00	15.00

Flying Saucers, #43-$20.

Flying Saucer Menace, The, published by Universal Publishing/Universal-Tandem Ltd. By Brad Steiger and August Roberts, one-shot, 90 photographs, 1967-$23.

Flying Saucers, #53-$18.

Flying Saucers, #38-$20.

Item	Good	Mint
Flying Saucers, published by Palmer Publications, Amherst, WI Editor, Helga Onan, 1950s-1970s		
Most 1950s issues	15.00	25.00
Most 1960s issues	10.00	20.00
Most 1970s issues	8.00	15.00
#38, "On Man in the Universe," digest size, October 1964	12.00	20.00
#43, "Mars, the Red Planet," digest size, August 1965	12.00	20.00
#53, "The CIA and the Little Green Men," August 1967	10.00	18.00
#54, "Myth, Reality and Flying Saucers, October 1967	10.00	18.00
#55, "Asteroid Icarus - What if it hit the Earth?," December 1967	10.00	18.00
Flying Saucers and UFOs 1968 (see *UFOs* 1968)		
Flying Saucers and UFOs 1969 (see *UFOs* 1969)		
Flying Saucers, UFO Reports, Dell Publishing Editor, Carmena Freeman, 1960s		
#2, yellow photo cover, "Seeing is Prickles, Pressure and Belief," 1967	12.00	18.00

Flying Saucers, #55-$18.

Flying Saucers, #54-$18.

Flying Saucers, UFO Reports, #2, yellow photo cover, "Seeing is Prickles, Pressure and Belief," 1967-$18.

Item	Good	Mint
Fortean Times, published by John Brown Publishing Ltd., London Editors, Bob Rickard, Paul Sieveking, 1973-present		
#1	25.00	50.00
Most 1970s issues	12.00	20.00
Most 1980s issues	8.00	15.00
Most 1990s issues	4.00	8.00
#75, June-July 1994 - UFO Cover-Up, Free Alien Defence Kit	5.00	8.00
#81, July 1995 - Saucers Buzz Scotland	5.00	8.00
#98, June 1997 - Happy Birthday UFO: 50 Years of Close Encounters	4.00	6.00
#109, May 1998 - Mexican Wave: Camcorder UFOs	4.00	6.00

Fortean Times, #75 -$8.

Fortean Times, #81 -$8.

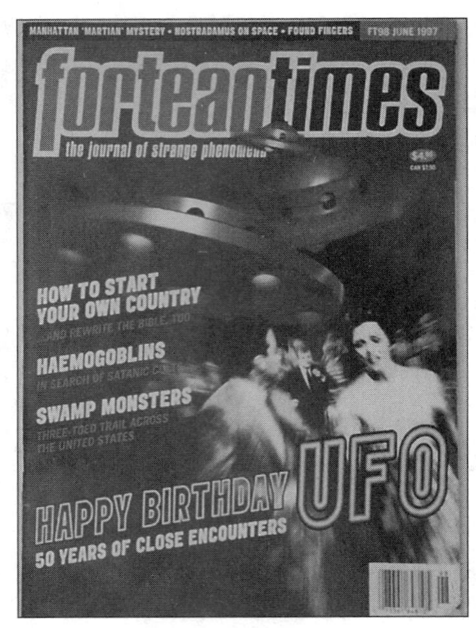

Fortean Times, #98 -$6.

Fortean Times, #109-$6.

Item	Good	Mint
Global UFO Investigation Magazine, published by Creative Force, England Editor, Andrew Hermidia, 1990s		
Most issues	5.00	10.00
#3, "The UK Exclusive," Oct./Nov. 1997	5.00	8.00
Ideal's UFO Magazine, published quarterly by Ideal Publishing Corp., New York Publisher, Phil Hirsch; Editor, D.C. Thrope 1970s		
Most issues	10.00	15.00
#1, "They Caused Us Trouble in Viet Nam," March 1978	12.00	18.00

Item	Good	Mint
International UFO Library magazine, published by The UFO Library, Studio City, CA Editor, Edward T. Foster 1990s (discontinued)		
Vol. 1, #4, "A Search for Truth," 1992	3.00	6.00
Vol. 3, #2, "USA, France, England, etc.," 1994	3.00	6.00
IUR (International UFO Reporter), published by the J. Allen Hynek Center for UFO Studies, Chicago Editor, Jerome Clark, 1976-present reprinted back issues available)		
Most issues	5.00	8.00
Vol. 19, #6 - November/December 1994	5.00	8.00

Global UFO Investigation Magazine, #3, "The UK Exclusive," Oct./Nov. 1997-$8.

International UFO Library magazine, Vol. 1, #4-$6.

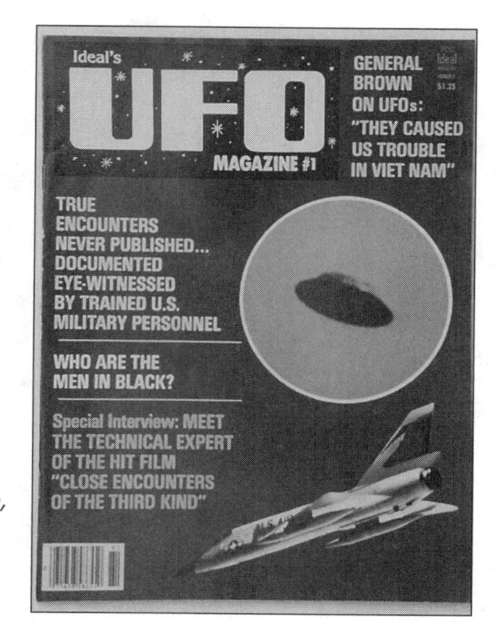

Ideal's UFO Magazine, #1, "They Caused Us Trouble in Viet Nam," March 1978-$18.

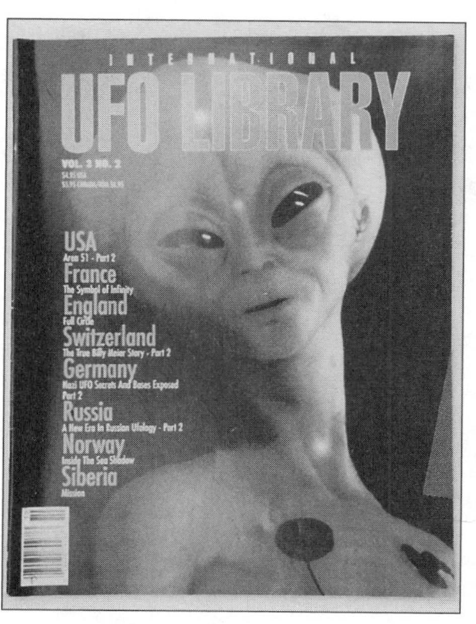

International UFO Library magazine, Vol. 3, #2-$6.

IUR (International UFO Reporter), Vol. 19, #6 - November/ December 1994- $8.

MUFON 1998 International UFO Symposium Proceedings-$30.

MAD, Super Special #117- $6.

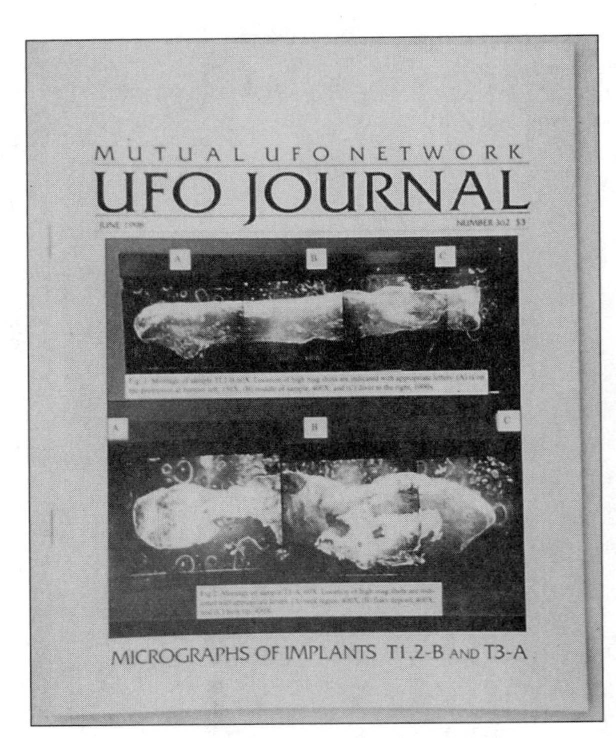

Mutual UFO Network—UFO Journal, #362,-$7.

Item	**Good**	**Mint**
MAD, Super Special #117, "E.T.s, Sci-Fi , etc.", E.C. Publications, New York, Dec. 1996	3.00	6.00
MUFON 1998 International UFO Symposium Proceedings, published by MUFON, Denver, CO Editors, Walter H. Andrus Jr. and Irena Scott, Ph.D., one-shot, 1998	20.00	30.00
Mutual UFO Network—UFO Journal, published by MUFON Editor, Dwight Connelly, 1960s-present (reprinted back issues available)		
Most issues	4.00	7.00
#362, June 1998 - Micrographs of Implants	4.00	7.00

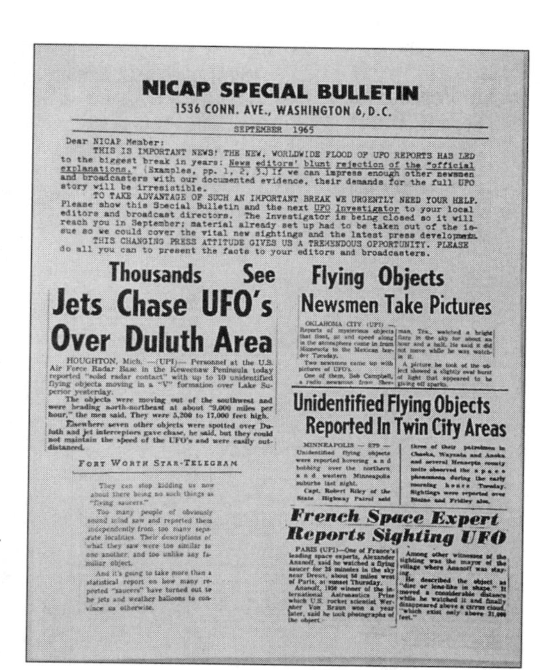

NICAP SPECIAL BULLETIN

1536 CONN. AVE., WASHINGTON 6, D.C.

SEPTEMBER 1965

Dear NICAP Member:

THIS IS IMPORTANT NEWS! THE NEW, WORLDWIDE FLOOD OF UFO REPORTS HAS LED to the biggest break in years: News editors' blunt rejection of the "official explanations." (Examples, pp. 1, 2, 3.) If we can impress enough other newsmen and broadcasters with our documented evidence, their demands for the full UFO story will be irresistible.

TO TAKE ADVANTAGE OF SUCH AN IMPORTANT BREAK WE URGENTLY NEED YOUR HELP. Please show this Special Bulletin and the next UFO Investigator to your local editors and broadcast directors. The Investigator is being closed so it will reach you in September; material already set up had to be taken out of the issue so we could cover the vital new sightings and the latest press developments.

THIS CHANGING PRESS ATTITUDE GIVES US A TREMENDOUS OPPORTUNITY. PLEASE do all you can to present the facts to your editors and broadcasters.

Thousands See Flying Objects

Jets Chase UFO's Over Duluth Area

HOUGHTON, Mich. —(UPI)— Personnel at the U.S. Air Force Radar Base in the Keweenaw Peninsula today reported "solid radar contact" with up to 10 unidentified flying objects moving in a "V" formation over Lake Superior yesterday.

The objects were moving out of the southwest and were heading north-northeast at about "9,000 miles per hour," the men said. They were 5,200 to 17,000 feet high.

Elsewhere seven other objects were spotted over Duluth and jet interceptors gave chase, he said, but they could not maintain the speed of the UFO's and were easily outdistanced.

FORT WORTH STAR-TELEGRAM

They can stop kidding us now about there being no such things as "flying saucers."

Too many people of obviously sound mind saw and reported them independently from too many separate localities. Their descriptions of what they saw were too similar to one another and too unlike any familiar object.

And it's going to take more than a statistical report on how many reported "saucers" have turned out to be jets and weather balloons to convince us otherwise.

Newsmen Take Pictures

OKLAHOMA CITY (UPI) — Reports of mysterious objects that float, sit and speed along in the atmosphere came in from Minnesota to the Mexican border Tuesday.

Two newsmen came up with pictures of UFO's.

One of them, Bob Campbell, a radio newsman from Sherman, Tex., watched a bright flare in the sky for about an hour and a half. He said it did not move while he was watching it.

A picture he took of the object showed a slightly oval burst of light that appeared to be giving off sparks.

Unidentified Flying Objects Reported In Twin City Areas

MINNEAPOLIS — ETP — Unidentified flying objects were reported hovering and bobbing over the northern and western Minneapolis suburbs last night.

Capt. Robert Riley of the State Highway Patrol said three of their patrolmen in Chaska, Wayzata and Anoka and several Hennepin county units observed the space phenomena during the early morning hours Tuesday. Sightings were reported over Blaine and Fridley also.

French Space Expert Reports Sighting UFO

PARIS (UPI)—One of France's leading space experts, Alexander Ananoff, said he watched a flying saucer for 35 minutes in the sky near Dreux, about 50 miles west of Paris, at sunset Thursday.

Ananoff, 1950 winner of the International Astronautics Prize which U.S. rocket scientist Wernher Von Braun won a year later, said he took photographs of the object.

Among other witnesses of the sighting was the mayor of the village where Ananoff was staying.

He described the object as "disc or lens-like in shape." It moved a considerable distance while he watched it and finally disappeared above a cirrus cloud, "which exist only above 21,000 feet."

NICAP Special Bulletin, September, 1965 issue-$25.

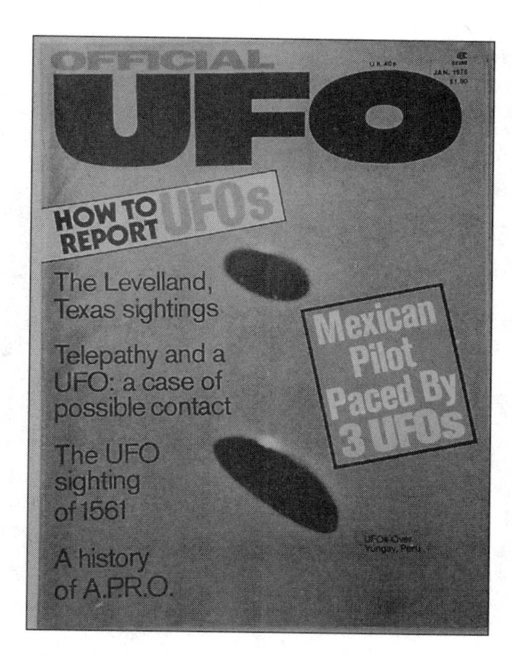

Official UFO, Vol. 1, #5-$15.

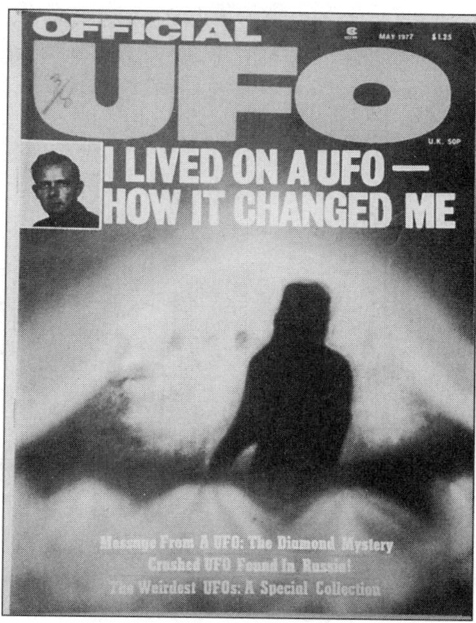

Official UFO, Vol. 2, #3-$15.

Official Guide to UFOs-$20.

Item	**Good**	**Mint**
Nexus (see *Saucer News*)		
NICAP Special Bulletin, published by NICAP, Washington D.C. (started by Donald Keyhoe)		
Editor (uncredited), 1950s-70s		
#1	25.00	35.00
Most 1950s issues	20.00	30.00
Most 1960s issues	15.00	25.00
Most 1970s issues	12.00	20.00
September, 1965 issue	15.00	25.00
Official Guide to UFOs, published by Science & Mechanics Publishing Co., New York		
Editor, John Scherer, one-shot, 1967	12.00	20.00

Item	**Good**	**Mint**
Official UFO, published by Countrywide Publications, New York		
Editor, Dennis William Hauck, mid-1970s		
#1	12.00	20.00
Most issues	8.00	18.00
Vol. 1, #5, "How to Report UFOs," Jan. 1976	8.00	15.00
Vol. 2, #3, 'I Lived On a UFO - How it Changed Me," May 1977	8.00	15.00
Vol. 3, #1, "Editor Jeff Goodman Kidnapped to Squelch Secret Information,"	8.00	15.00
Vol. 3, #3, "Illinois City Destroyed," April 1978	8.00	15.00
Official UFO Special: Ancient Astronauts, Jan. 1977	10.00	18.00
Official UFO Special: Ancient Astronauts, May 1977	10.00	18.00
Collector's Edition - Official UFO, Fall 1976	10.00	18.00

Official UFO, Vol. 3, #1-$15.

Official UFO, Vol. 3, #3-$15.

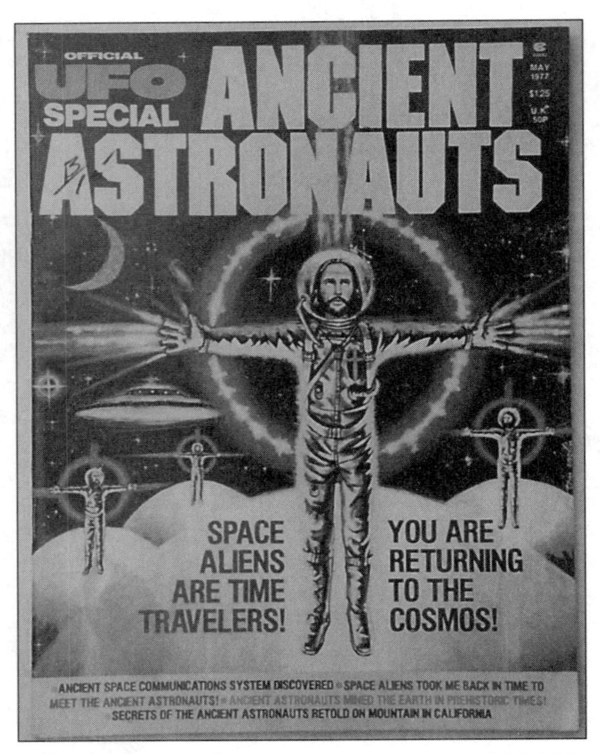

Official UFO Special: Ancient Astronauts, May 1977-$18.

Collector's Edition - Official UFO, Fall 1976-$18.

Item	Good	Mint
Saga's UFO Report (See *UFO Report*)		
Saucer News (first *Nexus*, much later *Saucer Smear*), published by the Saucer and Unexplained Celestial Events Research Society Editor, James W. Mosely, 1954-55 (*Nexus*), 1955-1972 (personal zine, typically traded, not sold)		
Under *Nexus* title:		
#1, July 1954	12.00	20.00
#2-5, 1954-55	10.00	15.00
#6, Adamski expose issue, Jan. 1955	12.00	20.00
#7-10, 1955	10.00	15.00
Under *Saucer News* title:		
Most 1950s issues	10.00	15.00
Most 1960s issues	8.00	12.00
Most 1970s issues	8.00	12.00
Saucer Smear (formerly *Saucer News, Nexus*), published by the Saucer and Unexplained Celestial Events Research Society Editor, James W. Moseley, 1981–present (personal zine, typically traded, not sold)		
Most issues	6.00	10.00
Vol. 35, #6, August 20, 1988	6.00	10.00
Saucers, published by Flying Saucers International Editor, Max B. Miller, 1953-1960		
Vol. I, #1, June 1953	20.00	25.00
Vol. III, #2, shows interiors of spaceships boarded by Adamski, June 1955	15.00	20.00
All other issues	12.00	20.00
Science & Mechanics, cover story, "...Most Awesome UFO Mystery," May 1967	3.00	6.00
Science & Mechanics, cover story, "Ithaca's Terrifying Flying Saucer Epidemic," July 1968	3.00	6.00
Strange Magazine, published by Mark Chorvinsky, Rockville, MD Editor, Mark Chorvinsky, 1987-present (two issues per year)		
Vol. 1, #1, Premier Double Issue, 1987	8.00	12.00
Most issues	5.00	8.00

Saucer Smear, Vol. 35, #6-$10.

Science & Mechanics, cover story, "...Most Awesome UFO Mystery," May 1967-$6.

Science & Mechanics, cover story, "Ithaca's Terrifying Flying Saucer Epidemic," July 1968-$6.

True Flying Saucers and UFOs Quarterly, #10-$12.

Item	Good	Mint
True Flying Saucers and UFOs Quarterly, published by Histrionic Publishing Co., New York Editor, Tom McArdell, 1970s		
Most issues	8.00	15.00
#10, "Terrifying East Coast Booms," Summer 1978	8.00	12.00
True Report on Flying Saucers, The, published by Fawcett Publications, Greenwich, CT Editor, Frank Bowers, 1960s		
#1, Exclusive Project Blue Book Sighting, 1967	12.00	20.00
#3, "Will the Astronauts Find Flying Saucers on the Moon?" 1969	12.00	20.00

True Flying Saucers and UFOs Quarterly, #10-$12.

True Report on Flying Saucers, The, #3-$20.

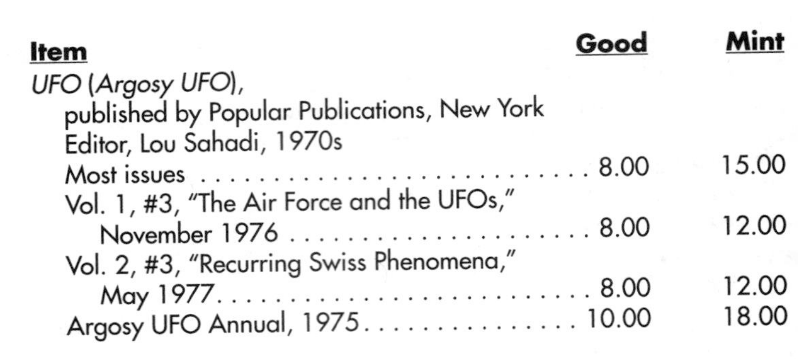

Item	Good	Mint
UFO (Argosy UFO), published by Popular Publications, New York Editor, Lou Sahadi, 1970s		
Most issues	8.00	15.00
Vol. 1, #3, "The Air Force and the UFOs," November 1976	8.00	12.00
Vol. 2, #3, "Recurring Swiss Phenomena," May 1977	8.00	12.00
Argosy UFO Annual, 1975	10.00	18.00

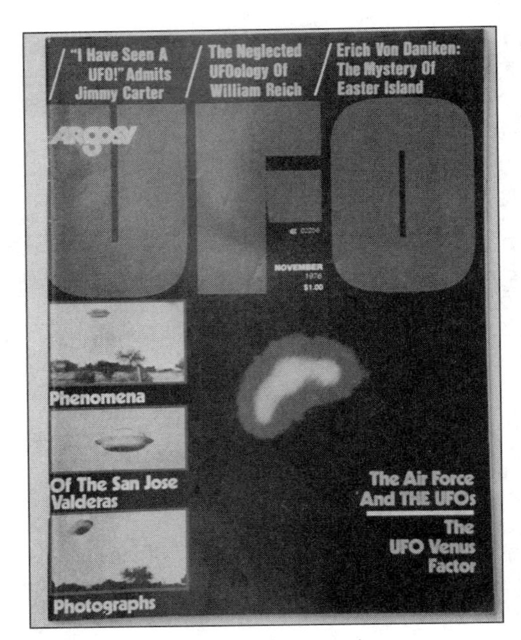

UFO (Argosy UFO), Vol. 1, #3-$12.

UFO (Argosy UFO), Vol. 2, #3-$12.

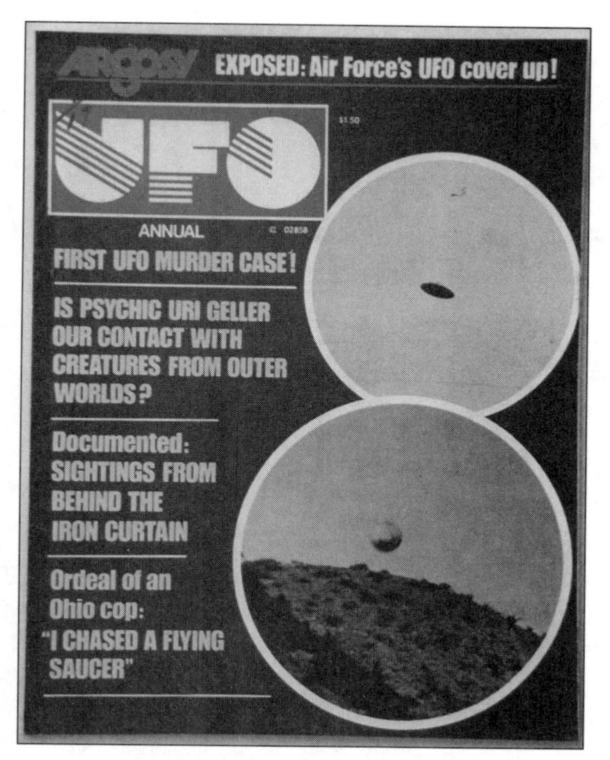

Argosy UFO Annual, 1975-$18.

UFO (California UFO, UFO Magazine), Vol. 2, #2-$6.

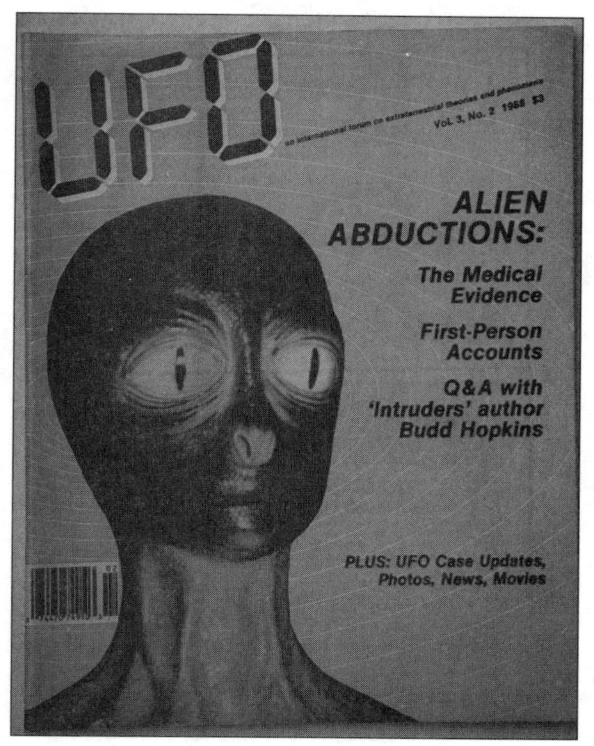

UFO (California UFO, UFO Magazine), Vol. 3, #2-$6.

UFO (California UFO, UFO Magazine), Vol. 3, #3-$6.

UFO (California UFO, UFO Magazine), Vol. 4, #3-$6.

UFO (California UFO, UFO Magazine), Vol. 6, #5-$5.

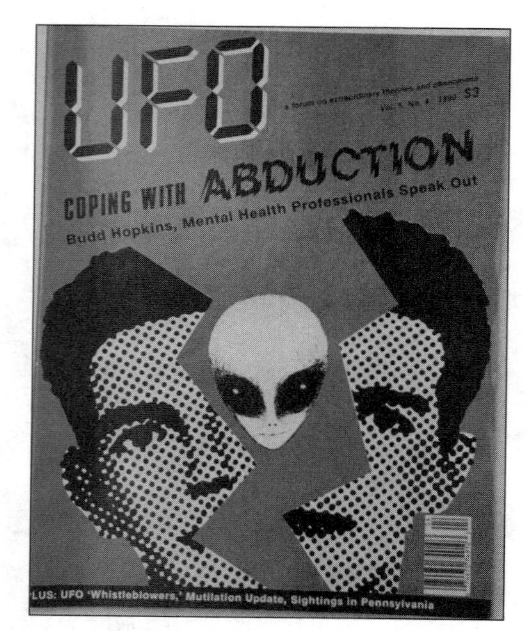

UFO (California UFO, UFO Magazine), Vol. 5, #4-$5.

UFO (California UFO, UFO Magazine), Vol. 8, #1-$5.

Item	Good	Mint
UFO (*California UFO, UFO Magazine*), published monthly, Los Angeles		
Editor, Vicki Cooper/Ecker 1986-present		
Vol. 1, #1	12.00	16.00
Most 1980s issues	10.00	15.00
Most 1990s issues (those unavailable from publisher)	10.00	15.00
Most 1990s issues (still available from publisher)	2.00	6.00
Vol. 2, #2, "Understanding Dolphins," 1987	3.00	6.00

Item	Good	Mint
Vol. 3, #2, "Alien Abductions," 1988	3.00	6.00
Vol. 3, #3, "UFOs in the News," 1988	3.00	6.00
Vol. 4, #3, "The Alien Question," July/August 1989	3.00	6.00
Vol. 5, #4, "Coping with Abduction," July/August 1990	2.00	5.00
Vol. 6, #5, "The Crop Circle Enigma," Sept./October 1991	2.00	5.00
Vol. 8, #1, "Encounter Research," Jan./Feb. 1993	2.00	5.00

Item	Good	Mint
Vol. 8, #2, "Mysteries of the Psyche," March/April 1993	2.00	5.00
Vol. 8, #3, "Fire in the Sky," May/June 1993	2.00	5.00
Vol. 9, #1, From Out of the Shadows," Jan./Feb. 1994	2.00	5.00
Vol. 9, #5, "Transforming Consciousness," Sept./Oct. 1994	2.00	5.00
Vol. 10, #1, "The UFO Elite," Jan./Feb/ 1995	2.00	5.00
Vol. 10, #3, "UFO Cults," May/June 1995	2.00	5.00
Vol. 11, #3, "Alien Implants," May/June 1996	2.00	5.00
Vol. 12, #2, "Alien Myths," March/April 1997	2.00	5.00

UFO (California UFO, UFO Magazine), Vol. 9, #1-$5.

UFO (California UFO, UFO Magazine), Vol. 8, #2-$5.

UFO (California UFO, UFO Magazine), Vol. 8, #3-$5.

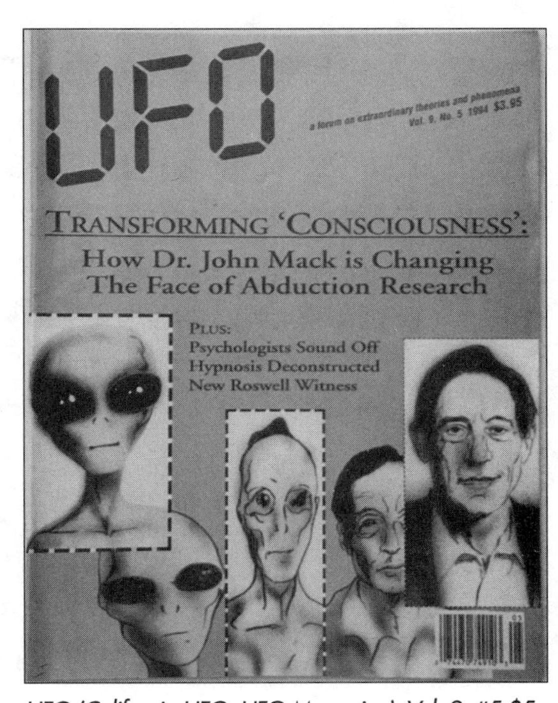

UFO (California UFO, UFO Magazine), Vol. 9, #5-$5.

UFO (California UFO, UFO Magazine), Vol. 10, #1-$5.

UFO (California UFO, UFO Magazine), Vol. 10, #3-$5.

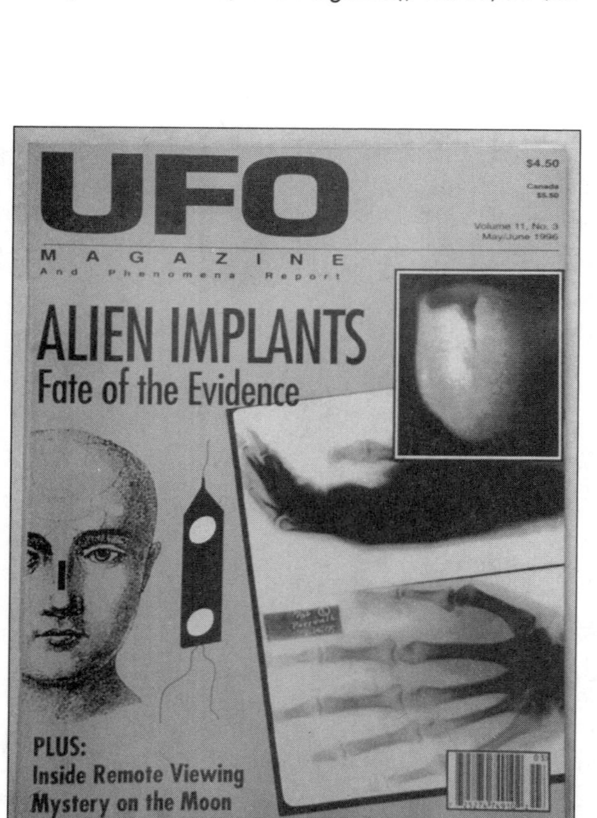

UFO (California UFO, UFO Magazine), Vol. 11, #3-$5.

UFO (California UFO, UFO Magazine), Vol. 12, #2-$5.

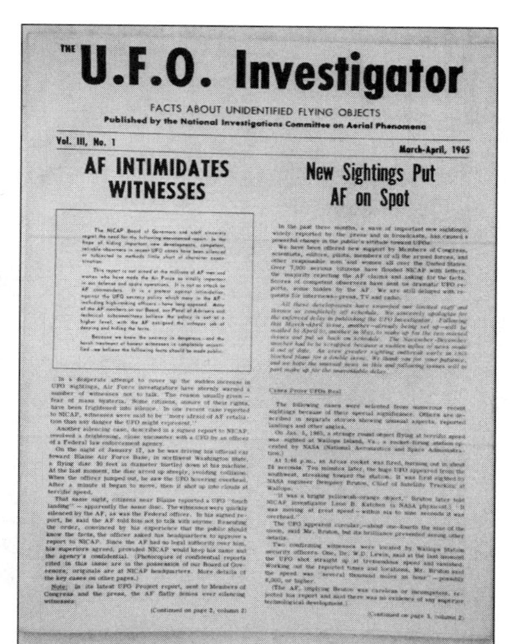
THE U.F.O. Investigator

FACTS ABOUT UNIDENTIFIED FLYING OBJECTS

Published by the National Investigations Committee on Aerial Phenomena

Vol. III, No. 1 — March-April, 1965

AF INTIMIDATES WITNESSES

New Sightings Put AF on Spot

U.F.O. Investigator, Vol. III, No. 1, March/April, 1965-$15.

UFO Magazine, Vol. 14, #1-$8.

UFO... Journal of Facts, Vol. III, Autumn 1991-$9.

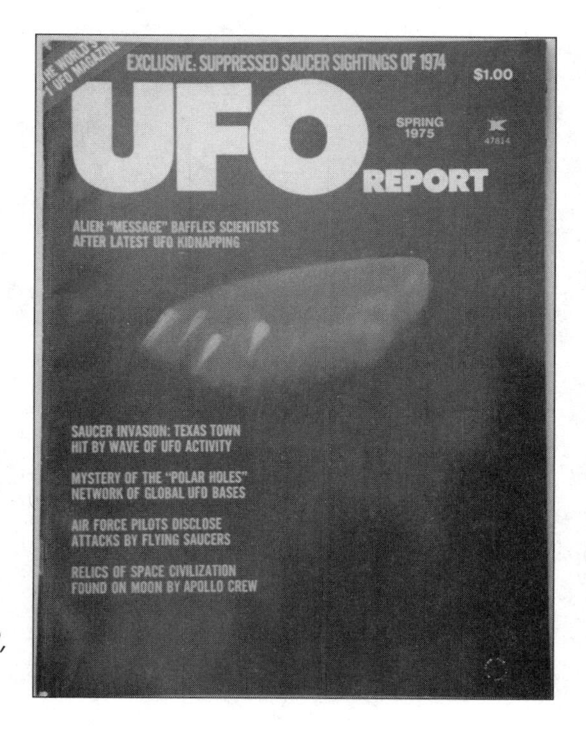

UFO Report (Saga's UFO Report), Vol. 2, #3-$12.

Item	Good	Mint
U.F.O. Investigator, published by the National Investigations Committee on Aerial Phenomena (NICAP club zine) Editor, Major Donald E. Keyhoe, 1950s-1970s		
Most 1950s issues	12.00	20.00
Most 1960s issues	10.00	15.00
Most 1970s issues	8.00	12.00
Vol. III, No. 1, March/April, 1965	10.00	15.00
UFO... Journal of Facts, published by UFO Photo Archives, Tucson Editor, Gem Gary Cox, 1980s-90s		
Most issues	5.00	10.00
Vol. III, Autumn 1991	6.00	9.00

Item	Good	Mint
UFO Magazine, published by Quest Publications Intl., England Editor, Graham William Birdsall, 1980s (70s?)-present		
Most pre-1990 issues	7.00	15.00
Most 1990s issues	4.00	8.00
Vol. 14, #1, "The Nashville UFO," May/June 1995	4.00	8.00

Item	Good	Mint
UFO Report (*Saga's UFO Report*), published by Gambi Publications, Brooklyn, New York Editor, David Elrich, 1970s		
Most issues	7.00	15.00
Vol. 2, #3, "Alien Message Baffles Scientists," Spring 1975	8.00	12.00
Vol. 4, #1, Jimmy Carter has seen UFO, May 1977	10.00	15.00
Vol. 4, #5, "Slaughter on the Prairies," September 1977	8.00	12.00
Vol. 7, #5, "Spitzbergen File," November 1979	7.00	10.00
UFO Review, Publisher/Editor, Timothy Green Beckley, newsprint fanzine, 1990s, each	5.00	10.00
UFO Sightings, published by S.J. Publications, Fort Lee, NJ Editor, Russell Wiener, 1980s		
Most issues	5.00	10.00
Vol. 2, #5, "CIA Supresses UFO Evidence," Sept. 1981	5.00	8.00

UFO Report (Saga's UFO Report), Vol. 4, #1-$15.

UFO Sightings, Vol. 2, #5-$8.

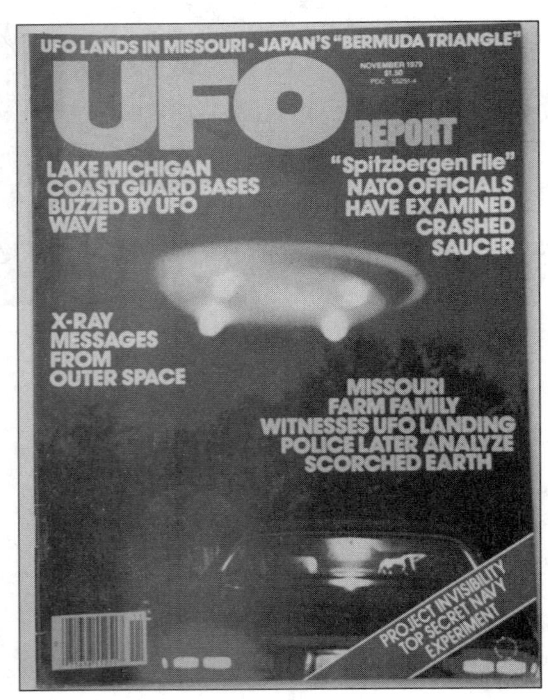

UFO Report (Saga's UFO Report), Vol. 7, #5-$10.

UFO Universe, Vol. 1, #4-$5.

UFO Universe, Vol. 1, #5-$5.

UFO Universe, Vol. 1, #5 (mismarked)-$5.

Item	Good	Mint
UFO Universe, published by GCR Publishing Group, New York Editor, Timothy Green Beckley, 1980s-90s		
Vol. 1, #1	4.00	8.00
Most issues	2.00	5.00
Vol. 1, #4, "Was Morton Downey Jr. Abducted by a UFO?" Winter 1989	3.00	5.00
Vol. 1, #5, "Beware! The Men in Black are Back," Summer 1989	3.00	5.00
Vol. 1, #5 (mismarked), "Are There Monsters on Mars?" January 1990	3.00	5.00
Vol. 2, #1, "Hidden Agenda Between Government and the Aliens," Spring 1992	2.00	5.00
Vol. 2, #4, "Super Scientists of Ancient Israel," Winter 1993	2.00	5.00
Vol. 3, #1, "UFO Crashes in New York City," Spring 1993	2.00	5.00
Vol. 5, #1, "Bizarre Aliens," Spring 1995	2.00	5.00
Vol. 5, #2, The X Archives," Summer 1995	2.00	5.00

UFO Universe, Vol. 2, #1-$5.

UFO Universe, Vol. 2, #4-$5.

UFO Universe, Vol. 3, #1-$5.

UFO Universe, Vol. 5, #1-$5.

UFO Universe, Vol. 5, #2-$5.

UFO Update, #11-$8.

Item	Good	Mint
UFO Update, published quarterly by eyond Reality Magazine Inc., New York Editor/Publisher, Harry Belil, 1970s-80s		
#1	8.00	12.00
Most issues	5.00	10.00
#11, "The Most Terrifying UFO Experience," Summer 1981	5.00	8.00
UFOs 1967, published by K.M.R. Publications, New York Editor, Tom McArdell, 1967		
#1	12.00	20.00

UFOs 1967, #1-$20.

UFOs 1968 (Flying Saucers and), #2-$20.

Unicus, Vol. 3, #2-$8.

Unicus, Vol. 4, #1-$6.

Item	Good	Mint
UFOs 1968 (*Flying Saucers and*), published by K.M.R. Publications, New York Editor, Tom McArdell, 1968		
#2, "Was Snippy Killed by Horse Butchers from Outer Space?"	12.00	20.00
UFOs 1969 (*Flying Saucers and*), published by K.M.R. Publications, New York Editor, Herbert M. Furlow, 1969		
#3	12.00	20.00
Uncensored UFO Reports, published by Goodman Media Group, New York Editor, Timothy Green Beckley, 1990s		
Most issues	3.00	6.00

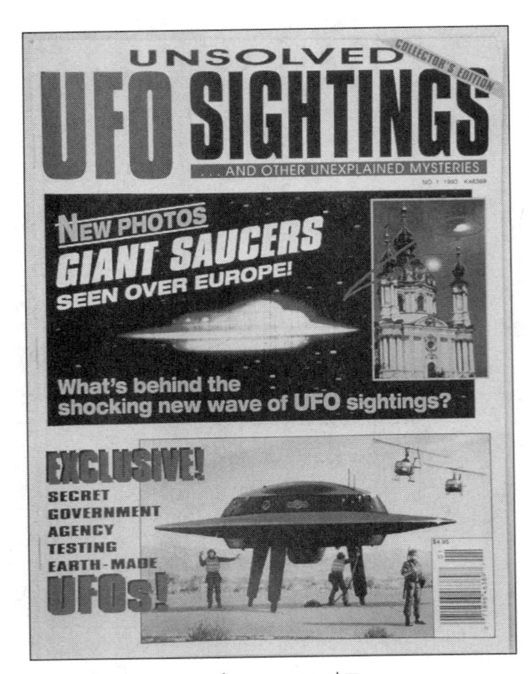

Unsolved UFO Sightings, #1-$7.

Unsolved UFO Sightings, #5-$6.

Item	Good	Mint
Unexplained Universe, published quarterly by GCR Publishing Group, New York Editor, Timothy Green Beckley, 1990s		
Most issues	3.00	6.00
Unicus (The magazine for Earthbound Extraterrestrials), Manhattan Beach, CA Editor, Robert M. Stanley, 1990s (discontinued)		
Most issues	4.00	10.00
Vol. 3, #2, "The E.T. Beat, Sacred Orgasms, What's New at Mu?" 1994	5.00	8.00
Vol. 4, #1, UFOs Over Malibu, 1995	4.00	6.00

Item	Good	Mint
Unsolved UFO Sightings, published by Charlotte Magazine Corp./ GCR Publishing Group, New York Editor, Timothy Green Beckley, 1990s		
Most issues	3.00	6.00
#1, "Giant Saucers Seen Over Europe," 1993	4.00	7.00
#2, "There is Life on Mars, 1993	3.00	6.00
#5, "Disc and Aircraft Battle It Out Over Florida," Fall 1994	3.00	6.00
Vol. 3, #4, "Nixon and the Roswell Alien Bodies," Winter 1996	3.00	6.00

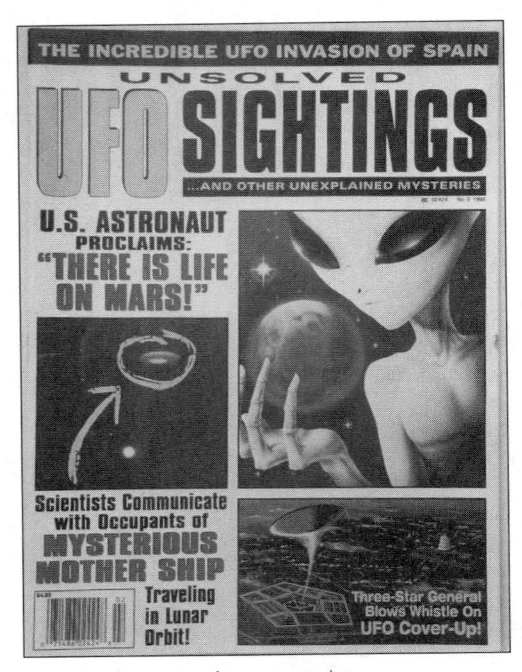

Unsolved UFO Sightings, #2-$6.

Unsolved UFO Sightings, Vol. 3, #4-$6.

Comic Books

Item	Good	Mint
Flying Saucers, Avon Periodicals/Realistic, 1950-1953		
#1, 1950, 21 pages	150.00	500.00
Reprint, not numbered, 1952, altered cover plus two extra pages not in original	100.00	300.00
Reprint, not numbered, 1953 edition	75.00	200.00
Flying Saucers, Dell Publishing, 1967-1969		
#1, April 1967	10.00	20.00
#2, July 1967	6.00	15.00
#3, October1967	5.00	12.00
#4, November 1967	5.00	12.00
#5, October 1969	5.00	10.00
Space Adventures, Charlton Comics, 1950s		
#4, Flying Saucer cover story	50.00	100.00
#6, Flying Saucer cover story	50.00	100.00
Space Adventures Presents... U.F.O., Charlton Comics, October 1967		
Vol. 3, #60	10.00	20.00
Strange Worlds, Marvel Comics, December 1958		
#1, Flying Saucer issue, Kirby and Ditko artwork	200.00	500.00
UFO Flying Saucers (became *UFO & Outer Space* with issue #14), Gold Key, 1968-1980		
#1, October 1968, 68 pages	10.00	25.00
#2, November 1970	6.00	12.00
#3, November 1972	6.00	12.00
#4, November 1974	6.00	12.00
#5, February 1975	5.00	10.00
#6-13, each	4.00	9.00
#14, title changes to UFO & Outer Space, reprints issue #3	3.00	6.00
#15-16, reprint issues, each	2.00	4.00
#17-25, each	2.00	5.00

Flying Saucers, Dell Publishing, #3-$12.

Space Adventures Presents... U.F.O., Charlton Comics, October 1967, Vol. 3, #60-$20.

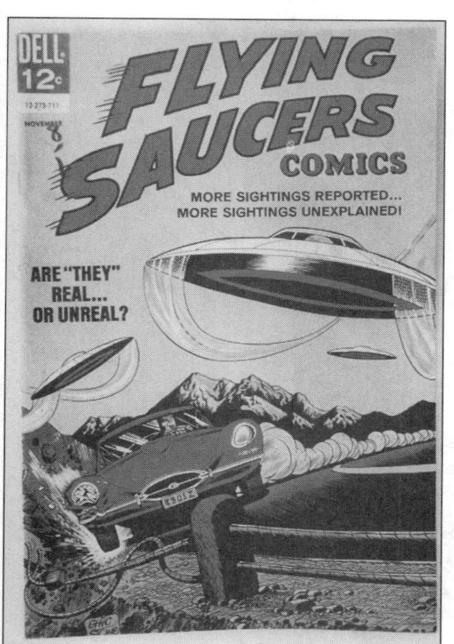

Flying Saucers, Dell Publishing, #4-$12.

UFO Flying Saucers, #1-$25.

Chapter 3
UFOs and Aliens in the Movies

20 Million Miles to Earth (1957)

An early Ray Harryhausen stop-motion animation classic, *20 Million Miles to Earth* tells the story of a Venusian that crash lands in the Mediterranean Sea and is taken to Italy. Once exposed to Earth's atmosphere, however, the tiny bipedal reptile, dubbed the Ymir, soon begins growing at an alarming rate. Eventually, it gets big enough to wrestle an elephant, and, as if that doesn't get the earthlings scared enough, it keeps on growing. Eventually, the Ymir is cornered at the Coliseum in Rome and killed.

Harryhausen's love of the King Kong legend shows in this story. The Ymir's only real crime is its size, and the only reason it becomes violent is because it is a victim of violence. This alien is one of the classics of the "monsters from space" genre made so popular in the 1950s, although his appearance is much more "Harryhausen" than the classic bug-eyed, big-brained monster. Harryhausen also created the special effects for *Earth vs. the Flying Saucers*.

Amazing Stories novel-$25.

Item	Good/Loose	Mint/MIP
Amazing Stories novel, digest format, 1957	15.00	25.00
Castle of Frankenstein magazine, Ymir cover, issue #20	12.00	20.00
Famous Monsters magazine, Ymir cover, issue #37, 1965	15.00	20.00
Model kit, spaceship, resin, Skyhook Models	35.00	50.00
Model kit, Ymir, by Billiken, resin, 1980s	550.00	650.00
Model kit, Ymir, vinyl, GEOmetric	40.00	70.00
Model kit, Ymir base, GEOmetric	40.00	70.00
Model kit, Diorama with Ymir bending lamp post, Classic Plastic	100.00	150.00
Model kit, Ymir battles elephant, Resin from the Grave	125.00	175.00
Model kit, super-deformed Ymir holding Dumbo, Mad lab Models	20.00	40.00
Monster Scene magazine, Ymir cover, issue #7	7.00	10.00
Monster Times magazine, Ymir cover, issue #20	8.00	12.00
Movie poster, one-sheet, 1957	200.00	300.00
Movie poster, three-sheet, 1957	250.00	400.00

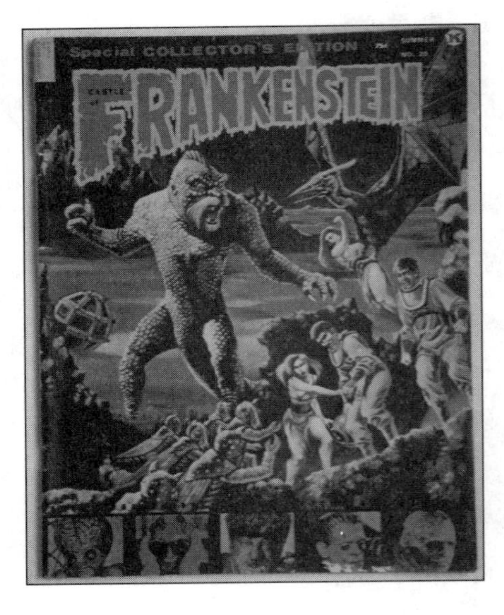

Castle of Frankenstein magazine-$20.

Famous Monsters magazine-$20.

The Blob (1958)

From the moment the Blob oozes onto the screen, viewers know this for sure—here is a new kind of alien design and alien threat. In the beginning, the Blob appears harmless. In fact, the old guy who first pokes at its meteorite shell with a stick doesn't even think it's alive—until, of course, it engulfs his hand and turns up the digestive juices. As the Blob gains size, appetite and momentum, it soon becomes clear to everyone that this alien life form does not come in peace. It seems, in fact, that, given the opportunity, it would devour every living creature on earth. Because bullets have no effect on its gelatinous body, the humans scramble to find a weakness. At last, with the aid of young Steve McQueen and his teen pals, a weapon is found—cold. The Blob can't stand cold. In the end, the Blob is pried off a classic 1950s diner and hoisted away to Antarctica. It was revived, however, in a pretty decent 1980s remake.

Item	Good/Loose	Mint/MIP
Model kit, 5 in., resin, Blob on diner, unmarked, Lunar Models	65.00	90.00
Movie poster, one-sheet, Paramount, 1958	150.00	250.00
Trading card, #101 from Horror Monster Series, Nu Card, 1961	5.00	10.00

Book, Close Encounters-$8.

Close Encounters of the Third Kind (1977)

Steven Spielberg's first foray into the "aliens on earth" genre took a much more adult approach than the later *E.T. Close Encounters of the Third Kind* tracks the journey of two people who have had contact with UFOs and been strangely changed by the experience. Drawn to recreate conical shapes, the contactees are eventually telepathically lured to Devil's Tower, a geological landmark in the Wyoming wilderness. Here, they witness the spectacular landing of a great alien mothership, containing friendly alien beings and loads of seemingly unharmed people who had been abducted over the years. The film was only loosely based on actual contactee reports and projects a mostly positive portrait of alien intervention in human affairs.

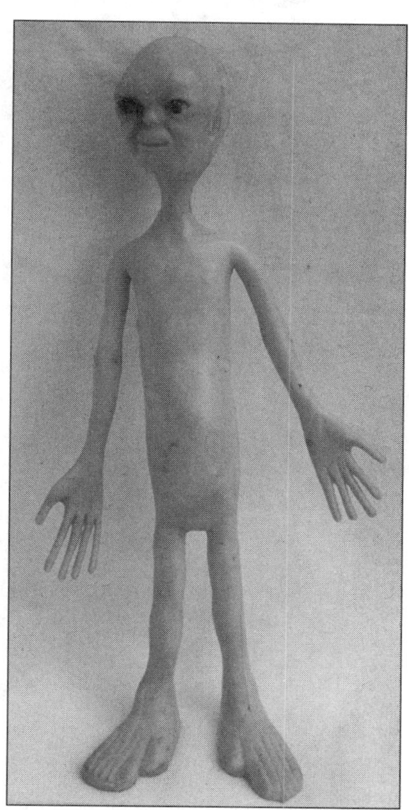
Alien bendee figure, 7"-$35.

Book, Close Encounters-$8.

Item	Good/Loose	Mint/MIP
Alien bendee figure, 7", Imperial/Columbia Pictures, 1977	15.00	35.00
Book, Close Encounters, Book Club Ed., hardcover, 1977	4.00	8.00
Cinefantastique, double issue, cover story, Vol. 7, No. 3-4	9.00	18.00

Fantastic Films magazine-$4.

Lobby cards (none show aliens or ships)-$8.

Item	Good/Loose	Mint/MIP
Comic book, Close Encounters..., Marvel Comics Super Special #3, 1978	2.00	5.00
Comic book, Close Encounters..., Marvel Special Edition #3, 1978	2.00	5.00
Famous Monsters magazine, cover story, issues #141, 1978	4.00	8.00

Item	Good/Loose	Mint/MIP
FantaScene magazine, cover story, #4, 1978	2.00	5.00
Fantastic Films magazine, cover story, April 1978	2.00	4.00
Fotonovel, 1977	8.00	15.00
Frisbee, blue plastic with decal, 1978	12.00	20.00
Game, Close Encounters, Parker Brothers, 1977-78	20.00	35.00
Lobby cards (none show aliens or ships), 11" x 14", 1977, each	6.00	8.00
Lunch box, metal, King Seely Thermos, 1978	30.00	50.00
Lunch kit thermos, King Seely Thermos, 1978	8.00	15.00

Magazine, Close Encounters-$8.

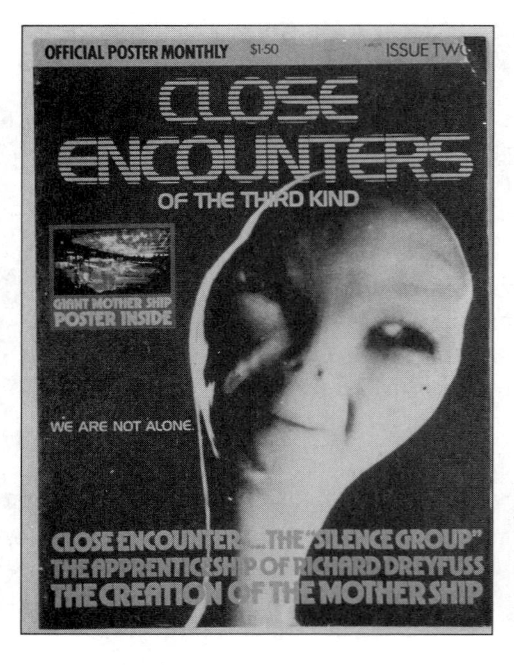

Magazine, Close Encounters Official Poster Monthly, issues 1-3, each-$12.

Movie poster, for Special Edition release, 1980s-$30.

Puzzle, jigsaw, "We Are Not Alone"-$18.

Science Fantasy Film Classics magazine-$4.

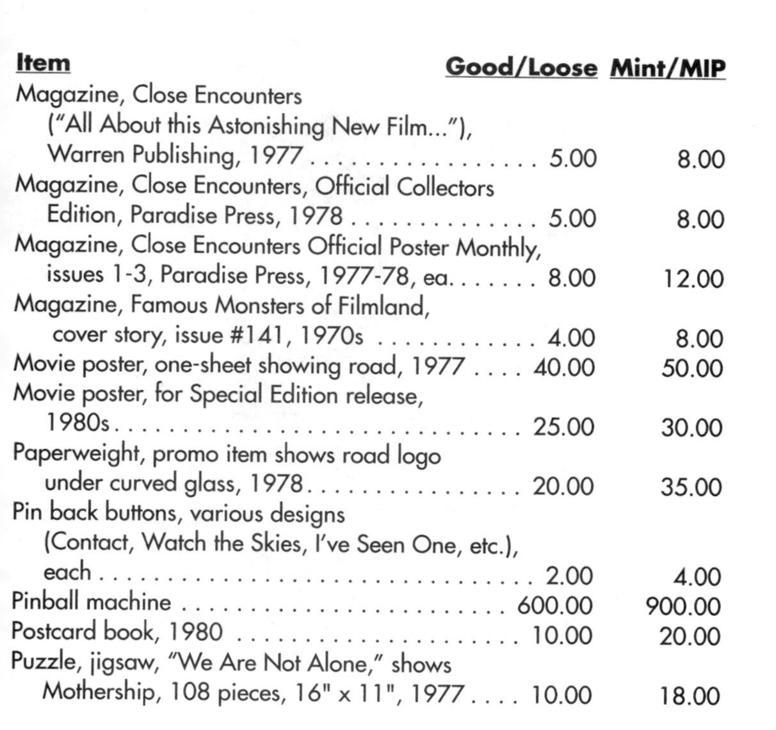

Item	Good/Loose	Mint/MIP
Magazine, Close Encounters ("All About this Astonishing New Film..."), Warren Publishing, 1977	5.00	8.00
Magazine, Close Encounters, Official Collectors Edition, Paradise Press, 1978	5.00	8.00
Magazine, Close Encounters Official Poster Monthly, issues 1-3, Paradise Press, 1977-78, ea.	8.00	12.00
Magazine, Famous Monsters of Filmland, cover story, issue #141, 1970s	4.00	8.00
Movie poster, one-sheet showing road, 1977	40.00	50.00
Movie poster, for Special Edition release, 1980s	25.00	30.00
Paperweight, promo item shows road logo under curved glass, 1978	20.00	35.00
Pin back buttons, various designs (Contact, Watch the Skies, I've Seen One, etc.), each	2.00	4.00
Pinball machine	600.00	900.00
Postcard book, 1980	10.00	20.00
Puzzle, jigsaw, "We Are Not Alone," shows Mothership, 108 pieces, 16" x 11", 1977	10.00	18.00

Starburst magazine-$3.

Starlog magazine-$10.

UFO magazine-$12.

Item	Good/Loose	Mint/MIP
Science Fantasy Film Classics magazine, cover story, Spring 1978	1.00	4.00
Starburst magazine, cover story, Vol. 1, #3, May 1978	1.00	3.00
Starlog magazine, cover story, #12, 1978	7.00	10.00
Starlog magazine, cover story, #38 (concurrent with Special Edition)	5.00	8.00
Trading cards, set of 66 cards, 11 stickers, Topps, 1978	12.00	20.00

Item	Good/Loose	Mint/MIP
Trading cards, Wonder Bread premium, set of 24, 1978	10.00	15.00
UFO magazine, cover story, 30th Anniversary Issue, 1978	7.00	12.00
UFO Report magazine, cover story, January 1978	1.00	3.00

Starlog magazine, cover story-$8.

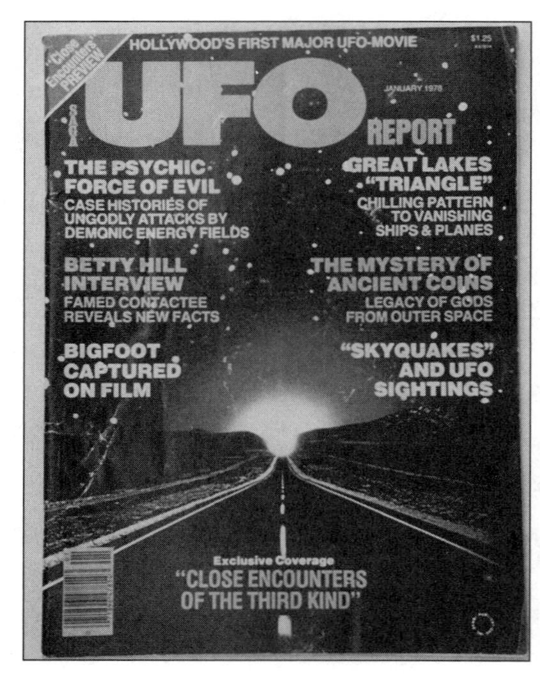

UFO Report magazine-$3.

Communion (1989)

Based on Whitley Strieber's national bestseller, *Communion* starred Christopher Walken as a man who encountered alien beings while in his cabin. Strieber claims that the film, directed by his friend, Phillippe Mora, is based on actual events, many of which are quite frightening.

Item	Good/Loose	Mint/MIP
Book, Communion: A True Story, Whitley Strieber, 1st edition, BeechTree Books, 1987	20.00	25.00
Book, Communion: A True Story, Whitley Strieber, Avon paperback, 1987	2.00	5.00
Movie poster, one-sheet, New Line, 1989	15.00	25.00

Movie poster-$25.

Critters (1986)

The original *Critters* film, along with its three sequels (released in 1986, 1988 and 1991) tells the story of a batch of fugitive fur-ball aliens who escape from their planet and land on earth, creating nothing but trouble and humorous sci-fi thrills. The special-effects people no doubt saved lots of money by simply throwing and rolling the vicious tribble-like alien critters toward their intended human prey.

Item	Good/Loose	Mint/MIP
Critters, model kit, Critter munching bone, 15 in., resin, unmarked	75.00	125.00
Critters, movie poster, one-sheet, New Line, 1986	15.00	25.00
Critters 2, movie poster one-sheet, New Line, 1988	12.00	20.00
Magazine, Fangoria, cover story, Critters 2, issue #74	7.00	10.00
Magazine, Monsterland (Forrest J. Ackerman's), issue #12, cover story, 1986	5.00	8.00

Movie poster-$50.

Day Mars Invaded Earth, The (1963)

A rocket scientist at Cape Canaveral helps to successfully deploy a communications device to Mars, then goes on vacation. When he comes home, he discovers evil Martian clones of himself and his family. The Martians set about killing the originals and reducing them to ashes. This film was released a full 10 years after *Invaders from Mars* and 7 years after *Invasion of the Body Snatchers*, both of which no doubt inspired its plot line.

Item	Good/Loose	Mint/MIP
Movie poster, one-sheet, 20th Century Fox, 1963	40.00	50.00

Day the Earth Stood Still (1951)

A classic of the "visiting aliens" genre, and one of the best science fiction films of all time, *The Day the Earth Stood Still* has Klaatu and his robot companion, Gort, arriving in Washington D.C. to warn earthlings about the dangers of war. We're told to stop using nuclear weapons or the more advanced space people will be forced to destroy our hostile planet. As a show of sincerity, Klaatu mysteriously interrupts all of the earth's power sources (except in hospitals and stuff) for 24 hours, giving the film its dramatic title.

Michael Rennie plays Klaatu, and Patricia Neal is the human female lead, forced at one point to utter the immortal words "Klaatu barada nikto," preventing the enormous Gort from engaging in a killing spree, starting with her. Gort was played by Lock Martin, a former employee of Grauman's Chinese Theater who just so happened to be about seven feet tall.

Item	Good/Loose	Mint/MIP
Magazine, Cinefantastique, cover shows Gort and Klaatu, Volume 4, No. 4, 1970s	10.00	20.00
Model kit, Gort and Klaatu set, Dimensional Designs	125.00	175.00
Model kit, includes space ship and two figures, Skyhook Models	45.00	65.00

Model kit-$60.

Earth vs. the Flying Saucers (1956)

Ray Harryhausen's whirling flying saucers highlight this paranoia-era classic. Viewers get to see most of Washington D.C. destroyed after White House bigwigs misunderstand an alien message. The aliens really did come in peace...we just read the note wrong and opened fire. Bad move on our part.

Item	Good/Loose	Mint/MIP
Model kit, 3-in. animating saucer, resin, Milestone Productions	40.00	60.00
Model kit, flying saucer, Skyhook Models	35.00	55.00
Model kit, Martian, Action Hobbies	75.00	100.00
Movie poster, one-sheet, Columbia, 1956	275.00	350.00
Movie poster, Australian daybill, 1950s	250.00	300.00

Lobby card-$80.

Flying Disc Man from Mars (1950)

Touted as "A Republic Serial in 12 Atomic Chapters," this story revives Mota the Martian (Gregory Gay) from *The Purple Monster Strikes*. This time he's armed with a "thermal disintegrator" and means to conquer earth from his volcano hideout. Can he be stopped? You bet.

Item	Good/Loose	Mint/MIP
Lobby card, title card, Republic Serial, 1950	60.00	80.00
Movie poster, one-sheet, Republic Serial, 1950	150.00	200.00

Lobby card-$80.

Flying Saucer (1950)

According to Michael "Psychotronic" Weldon, this is the first UFO movie ever made, and it required an official FBI viewing and thumbs-up before it was released to the public. Mikel Conrad produced, directed and stars in this one, about an agent sent to Alaska to investigate a UFO report. What he finds there are not aliens, but worse—Communists!

Item	Good/Loose	Mint/MIP
Lobby cards, 11" x 14", Film Classics, 1950, each	20.00	40.00

Item	Good/Loose	Mint/MIP
Lobby card, title card, 11" x 14", Film Classics, 1950	75.00	80.00
Movie poster, one-sheet, Film Classics, 1950	175.00	225.00

Hangar 18 (1980)

When a UFO is accidentally disabled by a government satellite, the US Military whisks the craft and its alien inhabitants into an airplane hangar, and a cover-up ensues. Darren McGavin and Robert Vaughn head the cast.

Item	Good/Loose	Mint/MIP
Movie poster, one-sheet, Sunn Classic Pictures, 1980	15.00	20.00

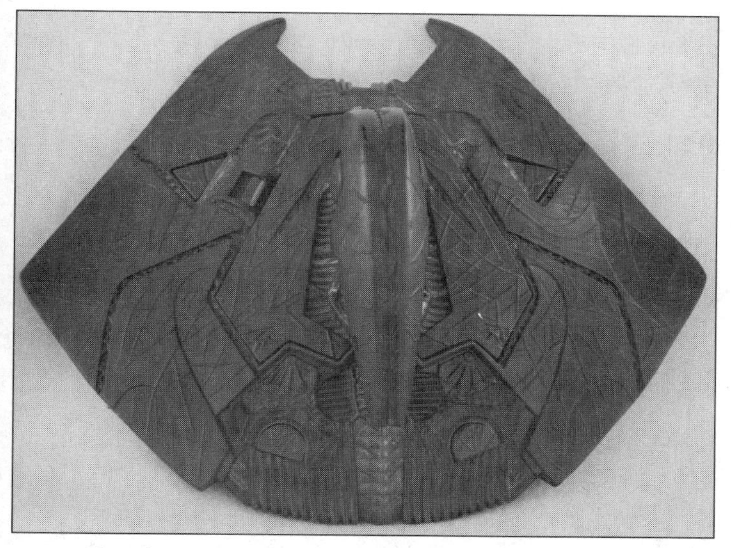

Space ship, from Independence Day, gray and blue plastic, 10-1/2" across, Trendmasters, 1996-$15.

Independence Day (1996)

Dean Devlin and Roland Emmerich scored a hit with this alien invasion epic. Will Smith is the hero of the day, a spirited soldier who fights the aliens in the air and on the ground. Brent ("Data") Spiner appears as a mad scientist character, studying the aliens in an underground Area 51-like setting. The first half of the film focuses on great shots of huge alien ships planting themselves above earth's major cities. Later, however, the film evolves into a patriotic, pro-war action banquet, as the good guys do stuff that's more and more unbelievable in each frame. Guess who wins.

Item	Good/Loose	Mint/MIP
Action figure, Alien Attacker Pilot, Trendmasters, 1996	6.00	10.00
Action figure, Alien Science Officer, Trendmasters, 1996	6.00	10.00
Action figure, Alien Shock Trooper, Trendmasters, 1996	6.00	10.00
Action figure, Alien Supreme Commander, Trendmasters, 1996	12.00	20.00
Action figure, Alien in Bio-Chamber (Suncoast exclusive), Trendmasters, 1996	12.00	20.00
Action figure, Capt. Steve Hiller, Trendmasters, 1996	6.00	10.00
Action figure, David Levinson, Trendmasters, 1996	6.00	10.00
Action figure, President Whitmore, Trendmasters, 1996	6.00	10.00
Action figure, Ultimate Alien Commander (FAO Schwartz exclusive), Trendmasters, 1996	20.00	30.00
Bio Chamber, action figure accessory, Trendmasters, 1996	3.00	8.00
Los Angeles Invasion playset, Trendmasters, 1996	20.00	30.00
Model kit, Alien Exoskeleton, Lindberg, 1996	12.00	16.00
Model kit, Captured Alien Attacker, Lindberg, 1996	12.00	16.00
Model kit, F/A-18 Hornet, Lindberg, 1996	12.00	16.00
Model kit, Stearman PT-17, Lindberg, 1996	12.00	16.00
Model kit, Alien with doctor, by Styrene Studio, 1996	150.00	200.00
Model kit, Alien Attacker, by Icons, 1996	800.00	1,000.00
Movie poster, one-sheet, 20th Century Fox, 1996	20.00	25.00
Movie poster, British quad size, 20th Century Fox, import, 1996	25.00	30.00
Space ship, gray and blue plastic, 10.5" across, Trendmasters, 1996	8.00	15.00
Trading cards, Topps, 1996		
individual card	15 cents	25 cents
full set (72 cards)	8.00	12.00
holo-foil card	4.00	6.00
unopened box	25.00	30.00

Invaders from Mars (1953 and 1986)

One of the earliest invasion genre classics, the original *Invaders from Mars* was directed by William Cameron Menzies, former children's illustrator who was also the set designer for *Gone with the Wind*. The movie, which has a dream-like atmosphere, is centered around a young boy who wakes from a nightmare to see a flying saucer land nearby; viewers follow the child's struggle for credibility. First, his father goes to investigate, only to come back strangely "transformed," taken over by alien forces. One by one, the boy's family and friends are enslaved by the Martians, until at last he convinces some adults to join the fight. Eventually, the boy and his doctor friend are captured by the Martians and taken to their underground space ship. This is where viewers get to see the Martian leader, a kid in shimmery make-up, a bubble head and tentacles sitting at the bottom of a glass orb. Wow. Good stuff. In the end, the army arrives, blows up the ship and the boy wakes up. Just when we think it really was a bad dream, he goes to his window and sees a saucer landing nearby...

Magazine, Filmfax-$100.

Item	Good/Loose	Mint/MIP
Magazine, Cinefantastique, cover story (Tobe Hooper remake), issue #60, 1986	3.00	5.00
Magazine, Fangoria, cover story (Tobe Hooper remake), issue #55, 1980s	4.00	7.00
Magazine, Filmfax, cover story, issue #2, 1986	50.00	100.00
Movie poster one-sheet, 20th Century Fox, 1953	400.00	600.00

Movie poster-$600.

Invasion of the Body Snatchers (1956 and 1978)

Invasion of the Body Snatchers tells a nightmarish tale of alien pods developing duplicates of sleeping humans, then killing and replacing them with emotionless, like-minded clones. Eventually, in the small town of Santa Mira, only two real humans remain. They are hunted mercilessly by the pod people and haunted by the knowledge that once they fall asleep, they will surely be killed and replaced. "You're next!" yells Kevin McCarthy, staring frantically out at the viewing audience, at the end of this classic. That ending was hacked out of some versions, replaced with a much happier resolution.

The 1978 remake fares very well, too, with Kevin McCarthy in the opening scene, reprising his final scene of the original. The remake also features Donald Sutherland, Brooke Adams, Leonard Nimoy, Veronica Cartwright and Jeff Goldblum, not to mention that incredibly disturbing pod-gone-awry creature—a dog with a man's face. Gotta love it.

Lobby card-$25.

Item	Good/Loose	Mint/MIP
Lobby card, Allied Artists, 1956 (scene dependent)	15.00	25.00
Movie poster, one-sheet, Allied Artists, 1956	350.00	550.00

Invasion of the Saucer Men (1957)

This one is so good. The aliens are perhaps the best-designed of all time, with huge brainy heads, gigantic bloodshot eyes and hands that are equipped with long hypodermic-type fingernails that inject their victims with booze. Their hands can also detach and grow eyes and maneuver about on their own. I defy you to find a cooler alien design anywhere.

This film, released a year before *The Blob*, takes a more lighthearted approach to the dilemma of aliens making first contact with teen-agers. Set in Hicksville, the film features drunken cows, Frank (The Riddler) Gorshin and Lyn Osborne (*Space Patrol's* Cadet Happy). Don't miss it.

Magazine, Famous Monsters of Filmland, cover story, issue #54-$10.

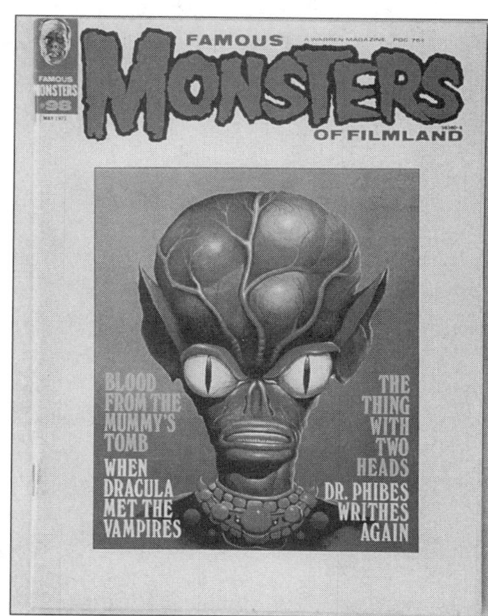

Magazine, Famous Monsters of Filmland, cover story, issue #98-$10.

Item	Good/Loose	Mint/MIP
Magazine, *Famous Monsters of Filmland*, cover story, issue #54, Warren Publishing, 1960s	5.00	10.00
Magazine, *Famous Monsters of Filmland*, cover story, issue #98, Warren Publishing, 1970s	5.00	10.00
Model kit, Invasion of the Saucer Men— Spaceship, by Skyhook Models, 1990s	50.00	75.00
Trading card, "Why Don't We Ever Have Dates," from Monster Laffs series, Topps, 1960s	75 cents	1.50

Trading card-$1.50.

It Came from Outer Space (1953)

Ray Bradbury wrote this story, and Jack Arnold (*Creature from the Black Lagoon*) directed it, so the film is somewhat classier than its title implies. An astronomer goes to check out a UFO he sees crash in the desert in this 3-D thriller. The aliens really only want to fix their ship and leave town, but they must duplicate earthlings to help make the needed repairs. This is supposedly the first film to depict aliens duplicating human townspeople, a feat destined to become a genre cliché.

Item	Good/Loose	Mint/MIP
Film, 8mm, 5.25" square box with generic sci-fi cover, Castle Films	15.00	25.00
View-Master preview reel, rare	150.00	250.00

Film, 8mm, 5-1/4" square box with generic sci-fi cover, Castle Films-$25.

It Conquered the World (1956)

In 1956, Roger Corman directed one of the greatest alien invasion B-films of all time —*It Conquered the World*. The legendary "carrot monster" has been immortalized in several resin and vinyl model kits. Corman tells the origins of its classic design:

"The *It Conquered the World* monster was designed to withstand the power of a strong gravity," he explains. "I was always interested in physics, so I wanted it to be done right. He was from a planet bigger than ours, so the gravity would be much stronger. So, I asked for a low, short monster to be designed.

"The first day on the set, Beverly Garland came in, and she walked up to the monster and stood over it, looking down at it, and said, 'So, you've come to conquer the world!'

"Well, I saw the point she was trying to make, and I realized that this was a good example of psychology overriding physics. So, I changed it. We broke for lunch, and I told the crew by the time I got back, that monster had to be 11 feet tall."

Item	Good/Loose	Mint/MIP
Magazine, *Fantastic Monsters of the Films*, cover story, issue #4, Black Shield Pubs., 1963	30.00	40.00
Magazine, *Filmfax*, cover story, issue #6, 1980s	25.00	50.00
Magazine, *Scary Monsters*, issue #1, cover story, Dennis Druktenis Publishing, 1991	7.00	10.00
Model kit, ("carrot monster") Billiken, 1980s	50.00	75.00
Movie poster, one-sheet, Sunset, 1956	100.00	175.00

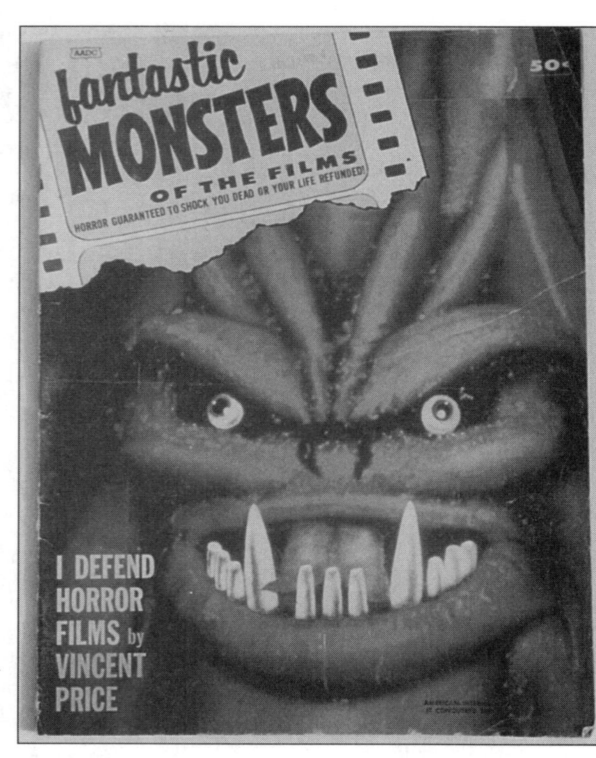

Magazine, Fantastic Monsters of the Films, cover story, issue #4-$40.

Mars Attacks (1996)

In 1996, Tim Burton released a film based on a series of gum cards from the early 1960s. The "Mars Attacks" cards, which are incredibly popular among collectors, horrified mothers at the time of their release, with wonderful, gory shots of big-brained Martians blasting humans, dogs and whatever crossed their path in a bloody, fiery invasion attempt.

The film remains true to the spirit of the original gum cards, as Burton offers up a visually stunning display of gore, humor and shocks. The film also includes a tribute to another '60s icon, Tom Jones, who not only sings, but becomes a character in the film, struggling to escape the aliens with a group of Las Vegas misfits.

Action figure, Martian Ambassador, electronic, 6"-$15.

Item	Good/Loose	Mint/MIP
Action figure, Doom Robot, 4", Trendmasters, 1996	3.00	8.00
Action figure, Doom Spider, 4", Trendmasters, 1996	3.00	8.00
Action figure, Martian Ambassador, 6", carded, Trendmasters, 1996	3.00	8.00
Action figure, Martian Ambassador, electronic, 6", carded, Trendmasters, 1996	5.00	15.00
Action figure, Martian Leader, 5", carded, Trendmasters, 1996	3.00	8.00
Action figure, Martian Leader, electronic, 5", carded, Trendmasters, 1996	5.00	15.00
Action figure, Martian Spy Girl, 5-1/2", carded, Trendmasters, 1996	20.00	40.00
Action figure, Martian Spy Girl, electronic, 5-1/2", carded, Trendmasters, 1996	30.00	50.00
Action figure, Martian Trooper, 4", Trendmasters, 1996	4.00	12.00

Action figure, Martian Spy Girl, electronic, 5-1/2"-$50.

Action figure, Martian Trooper, electronic, 5"-$15.

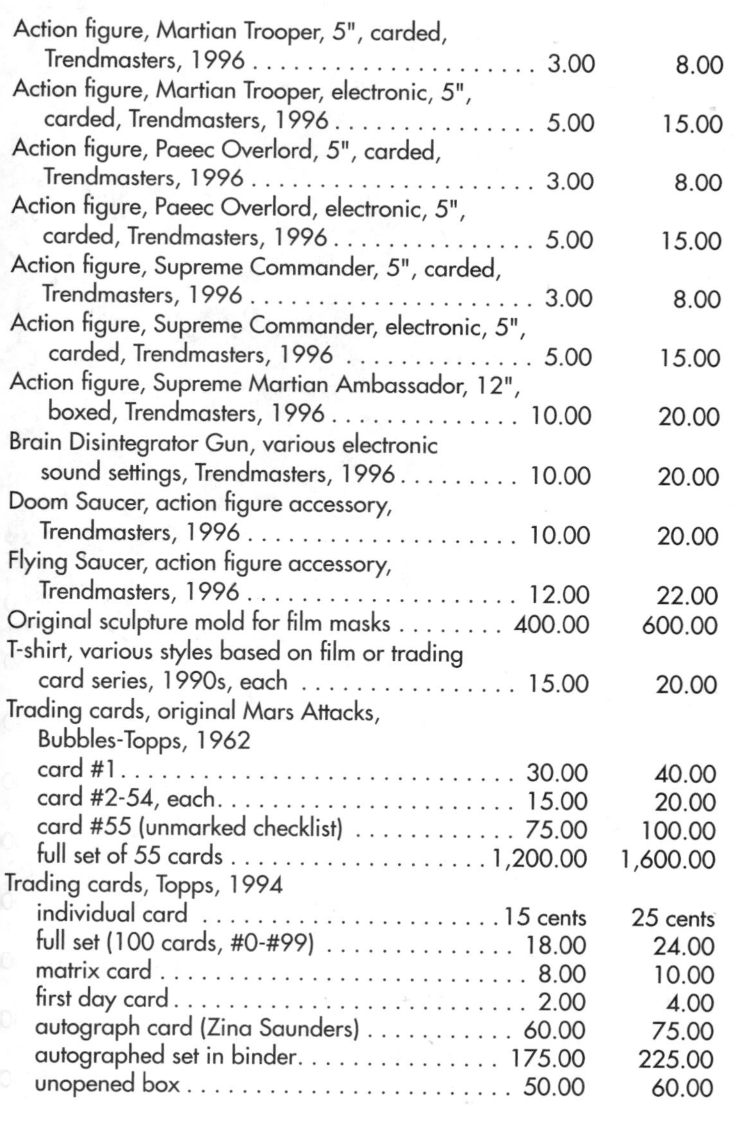

Action figure, Martian Trooper, 5", carded, Trendmasters, 1996	3.00	8.00
Action figure, Martian Trooper, electronic, 5", carded, Trendmasters, 1996	5.00	15.00
Action figure, Paeec Overlord, 5", carded, Trendmasters, 1996	3.00	8.00
Action figure, Paeec Overlord, electronic, 5", carded, Trendmasters, 1996	5.00	15.00
Action figure, Supreme Commander, 5", carded, Trendmasters, 1996	3.00	8.00
Action figure, Supreme Commander, electronic, 5", carded, Trendmasters, 1996	5.00	15.00
Action figure, Supreme Martian Ambassador, 12", boxed, Trendmasters, 1996	10.00	20.00
Brain Disintegrator Gun, various electronic sound settings, Trendmasters, 1996	10.00	20.00
Doom Saucer, action figure accessory, Trendmasters, 1996	10.00	20.00
Flying Saucer, action figure accessory, Trendmasters, 1996	12.00	22.00
Original sculpture mold for film masks	400.00	600.00
T-shirt, various styles based on film or trading card series, 1990s, each	15.00	20.00
Trading cards, original Mars Attacks, Bubbles-Topps, 1962		
card #1	30.00	40.00
card #2-54, each	15.00	20.00
card #55 (unmarked checklist)	75.00	100.00
full set of 55 cards	1,200.00	1,600.00
Trading cards, Topps, 1994		
individual card	15 cents	25 cents
full set (100 cards, #0-#99)	18.00	24.00
matrix card	8.00	10.00
first day card	2.00	4.00
autograph card (Zina Saunders)	60.00	75.00
autographed set in binder	175.00	225.00
unopened box	50.00	60.00

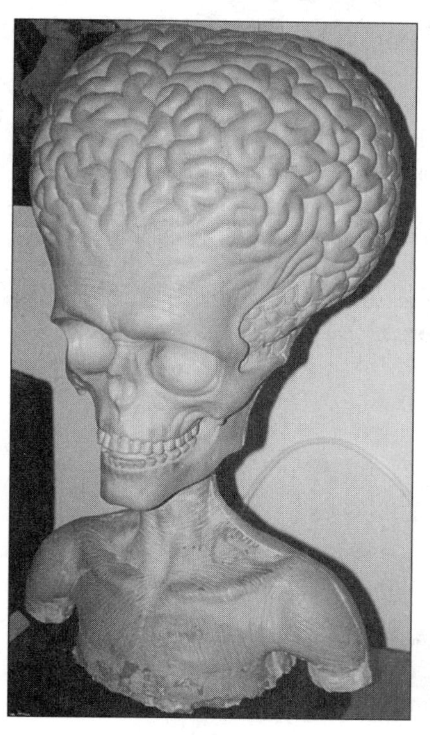

Original sculpture mold for film masks-$600.

Trading cards, Widevision version, Topps, 1996		
individual card	15 cents	25 cents
full set (72 cards)	12.00	15.00
Destruct-O-Rama card	6.00	8.00
unopened box	50.00	65.00
Trading cards, Mars Attacks Model Kits, Screamin', 1994 (Premium for purchasing all 8 model kits)		
individual card	35.00	50.00
full set (8 cards)	300.00	350.00
#9 bonus card	30.00	40.00
#10 5" x 7" card	30.00	40.00

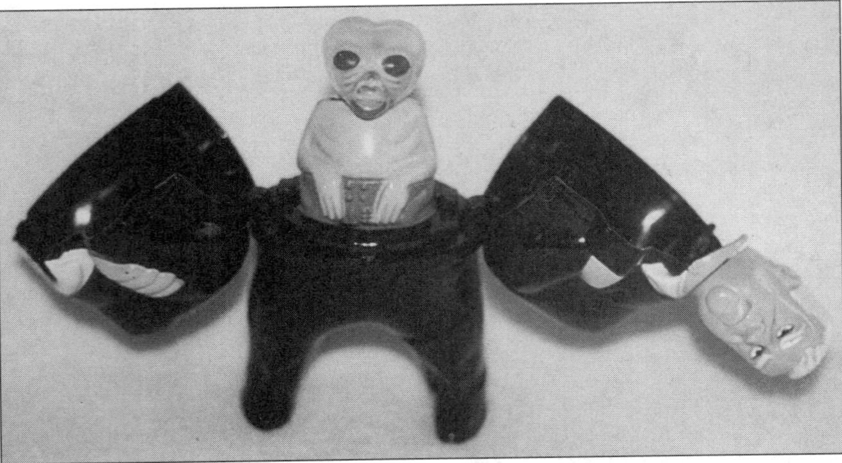

Fast food toy, Suited man opens to reveal alien, above, 4-1/4", Burger King, 1998-$4.

Men in Black (1997)

Will Smith and Tommy Lee Jones are the *Men in Black*, government liaisons and control agents for visiting alien entities. Armed with special anti-alien weaponry and flashing memory-erasers for the human contactees, the Men in Black protect the innocent. This stylish comedy action fest spawned a line of action figures, Burger King toys and a few other interesting collectibles.

Item	Good/Loose	Mint/MIP
Action figure, Alien Ambush Jay vs. Tree Stump Alien, 4", Galoob, 1997	5.00	10.00
Action figure, Alien Attack Edgar, 4", Galoob, 1997	5.00	10.00
Action figure, Alien Terrorist Edgar, Giant Aliens series, Galoob, 1997	12.00	20.00
Action figure, Flame Bustin' Jay vs. Scorched Alien, 4", Galoob, 1997	5.00	10.00
Action figure, Geebs, large alien series, Galoob, 1997	10.00	15.00
Action figure, Jay, large alien series, Galoob, 1997	10.00	15.00
Action figure, Mikey, large alien series, Galoob, 1997	10.00	15.00
Action figure, Slime Fightin' Kay vs. Edgar the Alien, 4", Galoob, 1997	5.00	10.00
Action figure, Street-Striker Kay vs. Manhole Alien, 4", Galoob, 1997	5.00	10.00
Bendee, Bobo the Squat, 4", Galoob, 1997	3.00	5.00
Bendee, Elby 17, 4", Galoob, 1997	3.00	5.00
Bendee, Mavis 12, 4", Galoob, 1997	3.00	5.00
Bendee, Mavis 13 and Seeble, 4", Galoob, 1997	3.00	5.00

Newsweek-$4.

Item	Good/Loose	Mint/MIP
Bendee, Mikey, 4", Galoob, 1997	3.00	5.00
Bendee, Neeble and Gleeble, 4", Galoob, 1997	3.00	5.00
Bendee, Redgick Jr., 4", Galoob, 1997	3.00	5.00
Bendee, Sulk, 4", Galoob, 1997	3.00	5.00
Bug Zapp 'em Van, boxed vehicle, action figure accessory, Galoob, 1997	12.00	20.00
Fast food toy, Suited man opens to reveal alien, 4-1/4", Burger King, 1998	2.00	4.00
Gun, Alien Blaster S-4 D-Atomizer, Galoob, 1997	20.00	30.00
Micro Machines, Collection 1, Galoob, 1997	5.00	10.00
Micro Machines, Collection 2, Galoob, 1997	5.00	10.00
Micro Machines, Collection 3, Galoob, 1997	5.00	10.00
Micro Machines, Collection 4, Galoob, 1997	5.00	10.00
Movie Poster, one-sheet, 1997	20.00	25.00

Item	Good/Loose	Mint/MIP
Newsweek, cover story, "Alien Busters," July 7, 1997	2.00	4.00
Trading cards, Inkworks, 1997		
individual card	15 cents	25 cents
full set (90 cards)	12.00	15.00
alien die-cut card	4.00	5.00
foilworks card	5.00	7.00
card in black	12.00	15.00
card album	15.00	20.00
unopened box	45.00	50.00

Phantom From Space (1953)

An invisible alien comes to earth and is trapped in the Griffith Observatory.

Item	Good/Loose	Mint/MIP
Movie poster, one-sheet, United Artists, 1953	150.00	200.00

Video-$15.

Plan 9 From Outer Space (1959)

Ed Wood's legendary alien invasion film, *Plan 9*, is touted by many as "The Worst Film Ever Made." This one features UFOs that are literally made of paper plates held by string. It also features the incomparable Tor Johnson, Vampira and Criswell. Bela Lugosi's initial appearance was spliced in following his death; his later scenes featured a stand-in shielding his face with a cape. Low on budget, big on campiness.

Item	Good/Loose	Mint/MIP
Magazine, *Filmfax*, cover story, issue #9, Tor and Vampira on cover, 1980s	10.00	20.00
Video, black and white, 79 minutes, Rhino, 1994	10.00	15.00

Predator (1987)

Arnold Schwarzenegger stars with the alien in the first *Predator* movie, playing on the popularity of both *Rambo* and *Alien*. Predator is a hunter from outer space; a perfect stalking and killing machine, who is after Arnold and his band of military buddies. He is hard to see and extremely merciless...the perfect fodder for a sequel film, a string of comic books and dozens of action figures.

Alien vs. Predator-$25.

Cracked Tusk Predator, 6"-$10, MIP

Action figures, Kenner, 1990s

Item	Good/Loose	Mint/MIP
Alien vs. Predator, 2-pack, 1993	15.00	25.00
Ambush Predator, mail-order only, clear plastic with silver, black accessories	20.00	30.00
Cracked Tusk Predator, 6", 1993	5.00	10.00
Lasershot Predator, electronic, 3 sounds, 6-1/2", 1995	12.00	20.00
Lava Planet Predator, 1995	5.00	10.00
Nightstorm Predator, 1994	5.00	10.00
Predator Blade Fighter	6.00	12.00

Predator Clan Leader, 7"-$12, MIP.

Ultimate Predators, white, left, and brown, right, each-$18, MIP.

Item	Good/Loose	Mint/MIP
Predator Clan Leader, 7 in, 1994	6.00	12.00
Scavage Predator, light green tint, with big, Y-shaped weapon, 5-1/2", 1993	15.00	25.00
Scavage Predator, darker green, 5-1/2", 1993	6.00	12.00
Spiked Tail Predator, 5-1/2", 1994	4.00	8.00
Stalker Predator, glows in the dark, 5-1/2", 1994	4.00	8.00
Ultimate Predator, large, 10" articulated figure, white, with smart disk weapon, 1995	10.00	18.00
Ultimate Predator, large 10" articulated figure, brown, with smart disk weapon, 1995	10.00	18.00
Other Items		
Collectors' Head, painted latex and foam	250.00	350.00
Comic book, *Dark Horse Comics* #1, Cover, Aug. '92	4.00	8.00
Comic book, *Predator*, by Dark Horse, issue #1	5.00	10.00
Comic book, *Predator*, by Dark Horse, issues #2-4, each	2.00	4.00
Comic book, *Predator—The Bloody Sands of Time*, #1 of 2	1.00	3.00
Comic book, *Predator—The Bloody Sands of Time*, #2 of 2	1.00	3.00
Comic book, *Predator vs. Magnus Robot Fighter* #1 of 2	2.00	5.00
Comic book, *Predator vs. Magnus Robot Fighter* #2 of 2	2.00	4.00
Fangoria magazine cover, issue # 65	12.00	20.00
Life-size Predator, foam latex standee, 8 ft. tall, Distortions, 1994	4,000.00	6,000.00
Mask, full head latex	50.00	75.00
Micro Machines, Collection 1, Galoob, 1996	4.00	7.00

Stalker Predator-$8, MIP.

Dark Horse Comics #1-$8.

Life-size Predator, foam latex standee-$6,000.

Model kit, Billiken-$75.

Item	Good/Loose	Mint/MIP
Micro Machines, Collection 2, Galoob, 1996	4.00	7.00
Micro Machines, Collection 3, Galoob, 1996	4.00	7.00
Micro Machines, Transforming Action Set, Galoob, 1996	6.00	12.00
Model kit, "Predator," by Gort, 7-1/2", pink resin, 4" x 6-1/2". base, unmarked	150.00	175.00

Item	Good/Loose	Mint/MIP
Model kit, "Predator 2," by Gort, 7-1/2", pink resin, oval base, unmarked	150.00	175.00
Model kit, Billiken, 12", soft vinyl kit, 1991	35.00	75.00
Model kit, cold cast porcelain, ltd. ed. of 1,000, 1/8 scale, Dark Horse	125.00	175.00
Movie poster, one-sheet, Predator, 20th Century Fox, 1987	20.00	35.00
Movie poster, one-sheet, Predator 2, 20th Century Fox, 1990	12.00	20.00
Trading cards, Fright Flicks wrapper with Predator portrait	50.00	100.00
TV Guide, Schwarzenegger cover (mentions Predator), April 21, 1990	5.00	10.00

Starship Invasions (1977)

This film is bizarre, even for the '70s. Robert Vaughn plays a UFO expert trying to figure out the alien agenda. Meanwhile, he is kidnapped by a group of good aliens, while another bad alien, played by Christopher Lee, plots to destroy the earth. Lee, as Captain Ramses, hits earth with his diabolical suicide ray, causing people to suddenly leap into traffic, strangle themselves or blow their brains out. Blame the Canadians for this one.

Item	Good/Loose	Mint/MIP
Lobby cards, 11" x 14", Warner Brothers, 1977, each	4.00	8.00
Movie poster, one-sheet, Warner Brothers, 1977	15.00	20.00

Lobby cards-$8.

This Island Earth (1955)

Although ridiculed in the recent *Mystery Science Theater 3000* feature film, *This Island Earth* remains a science fiction classic. The film's classic alien, the Metaluna Mutant (pronounced "mew-TANT") was a mid-1950s creation. And, while the Mutant's screen time was limited to a few minutes, its popularity is timeless.

Book-$40.

Item	**Good/Loose**	**Mint/MIP**
Book, *This Island Earth*, by R.F. Jones, Shasta hardcover	25.00	40.00
Halloween costume, Collegeville, 1980	50.00	75.00
Lobby card, This Island Earth, 1955 (scene dependent)	50.00	150.00
Model kit, Metaluna Mutant with Girl, Action Hobbies	100.00	150.00
Model kit, Metaluna Mutant, 1/5 scale, resin, Tsukuda #20, 1986	100.00	130.00

Movie poster one-sheet-$800.

Item	**Good/Loose**	**Mint/MI**
Monster Times, cover story, issue no. 9, 1970s	7.00	10.0
Movie poster one-sheet, This Island Earth, 1955	500.00	800.0
Mutant Raphael, TMN Turtle - Monster Turtles, Playmates, 1994	10.00	15.0
Nodder, model kit from Uncle Gilbert's	45.00	65.0
Poster, Basil Gogos ltd. ed. of 2,500, signed by artist, 1996	40.00	65.0

Nodder, model kit from Uncle Gilbert's-$65.

Poster, Basil Gogos ltd. ed. of 2,500, signed by artist, 1996-$65.

UFO (Unidentified Flying Objects) (1956)

Allegedly containing "actual color films of UFOs kept top secret until now," this documentary was narrated by Tom Powers.

Item	Good/Loose	Mint/MIP
Lobby card, 11" x 14", United Artists, 1956, each	8.00	10.00
Movie poster, one-sheet, United Artists, 1956	75.00	100.00
Movie poster, insert, United Artists, 1956	50.00	60.00

UFO Target Earth (1974)

A flying saucer is found in a lake in this low-budget '70s offering. Inside the craft are shape-shifting aliens composed of pure energy.

Item	Good/Loose	Mint/MIP
Movie poster, one-sheet, purple and white, Centrum, 1974	10.00	15.00

Lobby card, each-$10.

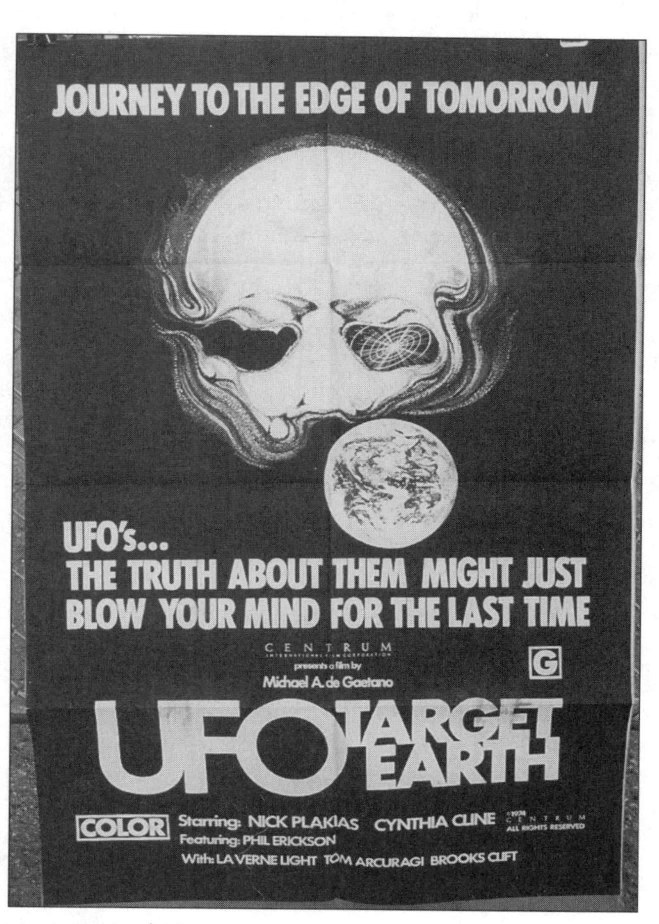

Movie Poster-$15.

War of the Worlds (1953)

Based on H.G. Wells' classic novel, this stunning film version was produced by George Pal. When Martians attack the earth, cities are destroyed and the military is powerless against them. In the end, it is earth's natural defenses, an organic virus, that stops the alien threat. The Martian warships are wonderful to watch, and an alien being, seen only for a second, is beautifully designed with one huge eye and long, thin arms.

Item	Good/Loose	Mint/MIP
Magazine, cover story, *Famous Monsters of Filmland*, issue #4, cover shows alien, 1950s	200.00	500.00
Model kit, Martian War Machine, L unar Models	100.00	150.00
Model kit, War of the Worlds Martian, Creature Features	100.00	150.00
Model kit, Martian War Machine, Skyhook Models	40.00	60.00
Model kit, Martian, Skyhook Models	45.00	65.00

Special Focus:
E.T. the Extraterrestrial

"What we relate to in E.T. is not what he looks like on the outside, but the goodness inside him."

-- Steven Spielberg
People Weekly, August 1982

In the summer of 1982, *E.T.* became the most successful movie of all time. Director Steven Spielberg claimed it was the first movie he'd made "for himself," and, at that time, had plans for a sequel. Today, more than 15 years after the film's release, fans have yet to see that sequel.

Instead, Spielberg went on to do other projects, his sequel ventures following *Raiders of the Lost Ark*, not *E.T.*

In fact, *E.T.* is quite an anomaly in the collectibles world. It's different because all the merchandise and all the long-lived popularity is based on one movie, alone. Typically, as with *Star Wars*, *Star Trek*, *Alien*, even *Ghostbusters* and *Gremlins*, science fiction collectibles have a broad base of connected "video." There is a series of movies, perhaps a TV show, or a Saturday morning tie-in...not so with *E.T.* He was a one-shot.

E.T. sprang from Spielberg's imagination one day as the director was shooting *Raiders of the Lost Ark* in the deserts of northern Africa. Separated from his girlfriend in Los Angeles and feeling lonely, he began pining for a special kind of friend: someone he could talk to, and someone who could give him "all the answers." Perhaps someone from very far away...

After discussing his initial ideas for the new movie with *Raiders* star Harrison Ford, Spielberg asked Ford's girlfriend, Melissa (*Black Stallion*) Mathison, to write a screenplay. Once shooting started on the project, Spielberg demanded secrecy, not wanting E.T.'s design to be imitated prematurely.

The movie opened to rave reviews, "stealing America's heart," according to the press. It tells the story of E.T., an extraterrestrial, who is left behind by his companions, stranded on earth, millions of light years from his home. He is found and sheltered by a young boy, Elliott (played by Henry Thomas). E.T. develops a strong bond with Elliott and his family (including baby sister played by 7-year-old Drew Barrymore). But, eventually, others discover the outsider, and government officials feel compelled to deal with the "alien presence" in their own way. Ripped from the innocent love of his young friends, and separated from an unnamed energy source of his home world, E.T. cannot survive the world of government-ordered tests and isolation. He begins to fade away, finally lapsing into a death state.

Miraculously, however, Elliott begins to sense that E.T.'s life force returns, and the young boy kidnaps his "corpse," leading cops and government thugs on a wild bicycle chase. Eventually, as E.T. regains full consciousness, the two evade their pursuers (re: famous flying bike against the full moon scene), and meet up with a beautiful spaceship from E.T.'s home world. We learn that it was the nearing presence of the ship that regenerated E.T., and we get to see him returning to his own people, preparing to go home.

People love the story (quite reminiscent of the *New Testament*), and E.T.'s fan base has survived for 15 years. E.T. collectors are still found in surprising numbers, dedicated fans of the lovable, lost alien.

Linda Stanley, a professor at Colorado State University in Fort Collins, began collecting E.T. in 1995. "I got my first E.T. stuff at a garage sale," she recalls. My husband (former mayor of Fort Collins) collects everything, so I started going to garage sales with him. Then, the movie aired on network television, and I thought, 'Wow, E.T. is such a cool guy, such a good spirit, such a gentle spirit...I'm gonna collect E.T.!' "

Today, Linda has nearly 450 pieces in her ever-growing collection. Her favorite is the elusive E.T. scooter, a Salvation Army find. The plastic scooter, scaled to toddler size, features an embossed image of E.T. in the bicycle basket on its front. Tops on her want list is the life-size E.T. replica produced by Sharper Image, which sells for a cool $1,200.

Most E.T. merchandise was produced in 1982, the year of the film's initial release. Six years later, in 1988, the video was released, with Pepsi and Sears sponsoring enthusiastic E.T. promotions and pushing the creature once again into the limelight.

Among the most popular E.T. collectibles today are the figures and plush toys produced in the early 1980s. LJN produced a series of figures, including the E.T. and Elliott Powered Bicycle, an E.T. wind-up walker, a Talking E.T., and a series of small 2-1/2" figures, featuring E.T. in different poses and outfits. Popular plush toys include a large (18") E.T., a smaller leatherette E.T. (Kamar, 1982) and a pink-ish plush E.T. produced by Applause in 1988.

Our pick for the coolest of the cool E.T. collectibles, however, is Knickerbocker's E.T. Finger Light. A replica of E.T.'s own long finger, fans can proudly wear the appendage, and when it presses against something, it lights up...just like in the movie! Now that's hip!

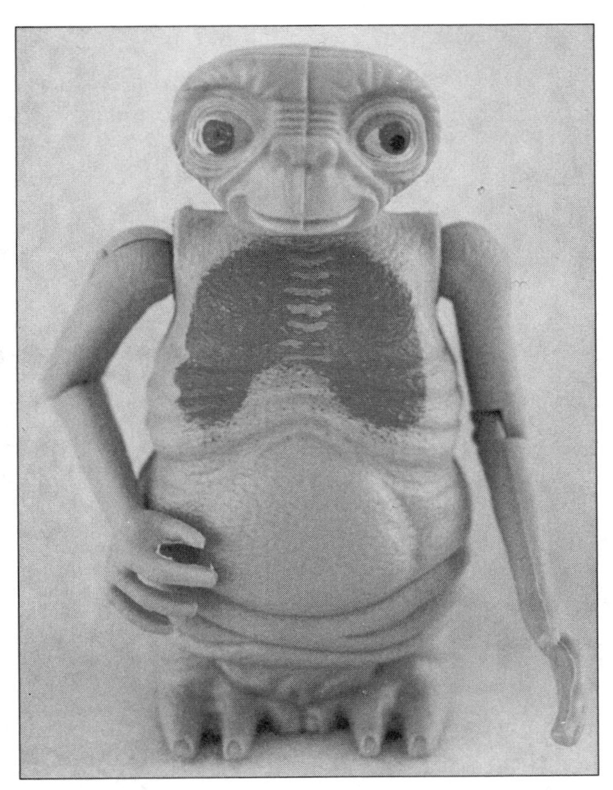

Action Figure, can move head up and down, jointed arms-$18, MIP.

Action figure, brown plastic-$20, MIP.

Atari E.T. game, 1982-$10.

Backpack-$20.

Bank, ceramic-$15, MIP.

Universal Studios E.T. bank-$12, MIP.

Bath set-$50.

E.T. Storybook-$12.

Item	Good/Loose	Mint/MIP
Action Figure, can move head up and down, jointed arms, 3-3/4", LJN, 1982	10.00	18.00
Action Figure, dressed in scarf, robe or dress and hat, LJN, 1982	10.00	18.00
Action Figure, glowing heart, LJN, 1982	10.00	18.00
Action figure, brown plastic, 6", made in Taiwan (unlicensed?)	12.00	20.00
Address Book, E.T. portrait on cover, 1982	12.00	20.00
Atari E.T. game, 1982	5.00	10.00
Backpack, blue fabric with graphic, E.T. Luggage tag, 16", Star Power, 1982	12.00	20.00
Balloons, pack with E.T. picture on front, 1982	4.00	8.00
Bank, ceramic, green base, squat, wrinkled E.T., 6", no mark	8.00	15.00
Bank, Universal Studios E.T. bank, round with E.T. head in middle, 7", U.C.S., 1982	8.00	12.00
Bath set, towel and wash mitt, towel is 22" x 42", Barth & Dreyfuss, 1980s	30.00	50.00
Book, E.T. Storybook, hardcover with color photos, 8-1/2" x 11", G.P. Putnam's Sons, 1982	8.00	12.00
Book and Record Set, Gertie and E.T. on cover, 1982	8.00	12.00
Bubble Bath container, figural E.T. with blue robe, 6", boxed, Avon, 1982	15.00	25.00
Calendar, E.T. the Extra-Terrestrial-1983, 11" x 12", Perigee Books, 1982	15.00	22.00
Candle, wax with glassy eyes, 6-1/2", Zooka Corp., 1982	10.00	18.00

Calendar-$22.

Bubble Bath container-$25, MIP.

Candle-$18, MIP.

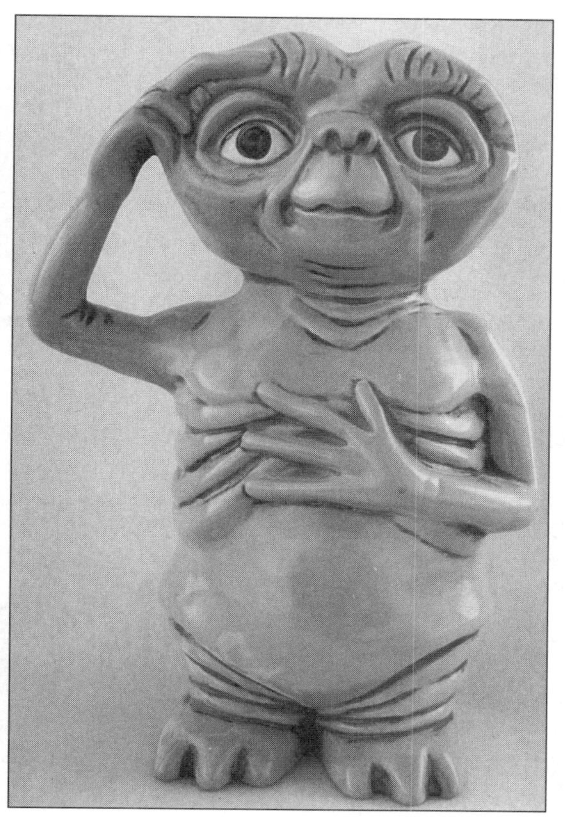
Ceramic figure-$20, MIP.

Ceramic "hang-on" decoration-$8.

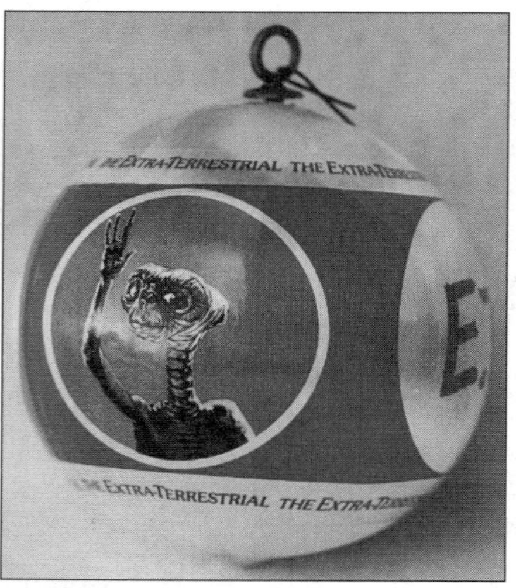

Christmas Ornament-$15.

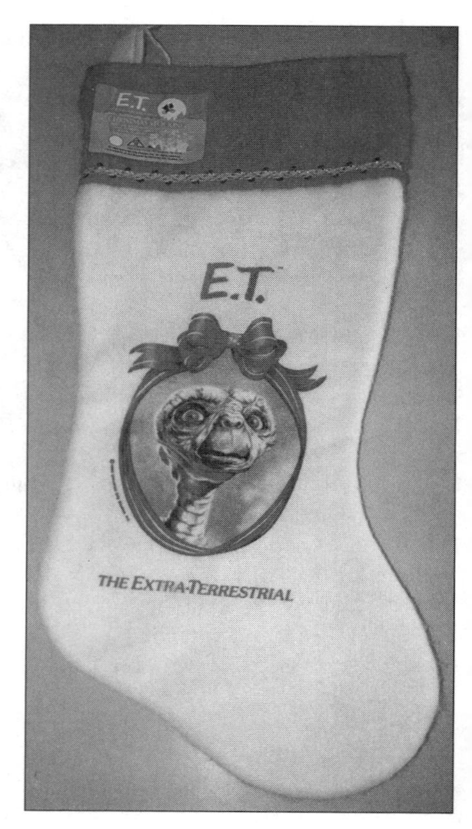

Christmas Stocking-$25.

Item	Good/Loose	Mint/MIP
Ceramic figure, hand made from mold, 9-1/2", 1982	10.00	20.00
Ceramic "hang-on" decoration, 3", Avon, 1983	5.00	8.00
Cereal Box, E.T. Cereal, General Mills, 1982	10.00	15.00
Christmas Ornament, round with decal, no mark, 1982	10.00	15.00
Christmas Stocking, felt with various designs, paper sticker, 1982	15.00	25.00
Clip-on E.T., furry brown body, 4", no mark	8.00	12.00
Cinefex magazine, cover story, issue #11, 1982	10.00	20.00
Colorforms, large box, E.T. illustration on front, 1982	20.00	35.00
Coloring Set, 3 large pictures to color, with felt tip pens, Fun Art, 1982	8.00	12.00
Costume, vinyl with light plastic string mask, Collegeville, 1982	15.00	30.00
Cup, ceramic, E.T. with flower pot, speckled background, no mark	5.00	10.00
Cup, plastic drinking cup with E.T. graphics, handle, 3-1/2", Deka Plastics, 1982	4.00	8.00
Cup topper, figural (straw goes in back), E.T. with flowers, 7", UCS, 1980s	5.00	10.00
Diecast E.T. holds up finder, 4-1/2", no mark	12.00	20.00
Diecast E.T. dressed in girl clothes, 3-1/2", no mark	12.00	20.00
Dinnerware, plastic 20-oz. bowl and cup set, Deka Plastics, 1982	10.00	18.00
Dinnerware, plastic plate, 8", E.T. with children, U.C.S. 1982	4.00	8.00
Drinking Glass, any design of four-glass set, 6", Pizza Hut, 1982, each (Gertie kisses E.T., E.T. with stuffed toys, E.T. with Elliott, and "Phone Home")	5.00	9.00
Drinking Glass, any one of four by AAFES/Paramount, 1982, each	8.00	12.00

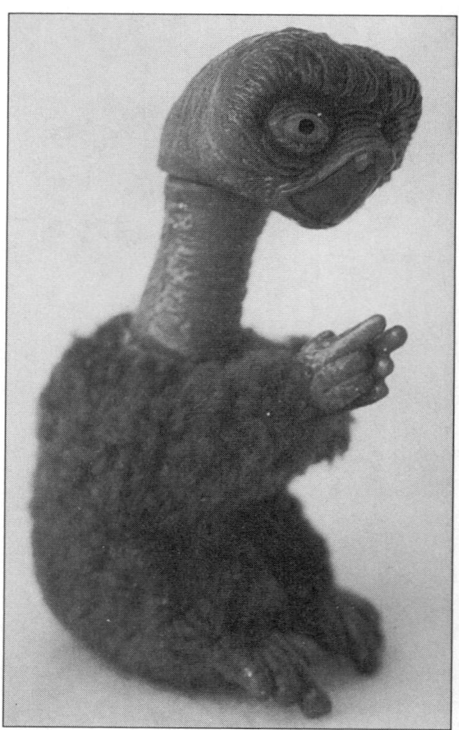

Clip-on E.T.-$12.

Item	Good/Loose	Mint/MIP
(E.T. in bike basket, flying past moon, with Gertie in costume, standing with Elliott)		
Eraser, Jumbo Eraser shaped like E.T., 4", on card, Star Power, 1982	3.00	5.00
E.T. and Elliott Powered Bicycle, on card, LJN, 1982	12.00	20.00

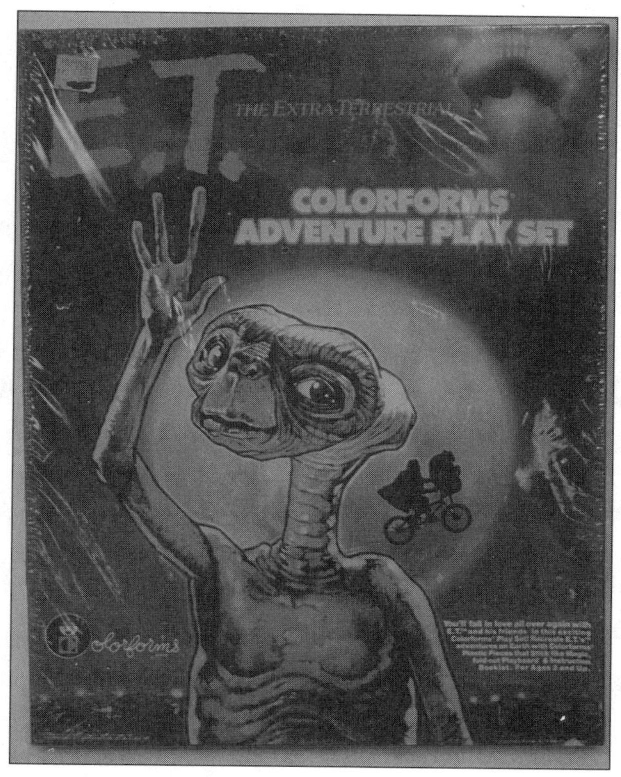

Colorforms-$35.

Costume-$30.

Cup, ceramic-$10.

Plastic drinking cup-$8.

Item	Good/Loose	Mint/MIP
E.T. and Spaceship Launcher, 3" tall, on card, LJN, 1982	15.00	25.00
E.T. Pop-up Spaceship, on card, LJN, 1982	12.00	20.00
Famous Monsters of Filmland magazine, cover story, issue #189, Warren Publishing, 1982	6.00	12.00
Famous Monsters of Filmland magazine, cover story, 1983 Yearbook, Warren Publishing	8.00	15.00
Fan Club application, shows E.T. peeking around door, 1982	1.00	2.00
Fan Club membership certificate, 1982-83	3.00	5.00
Fan Club membership card	6.00	10.00
Figure, 2-1/2", Assortment A, any of 6 figures, LJN, 1982, each	6.00	12.00
(Assortment A includes these figures: reading book, pointing with phone, blanket and Speak & Spell, in robe with beer, holding plant, dressed in drag)		
Figure, 2-1/2", Assortment B, any of 6 figures, LJN, 1982, each	6.00	12.00
(Assortment B includes these figures: hugging doll, with robe and phone, sheet over head, lifting flower in left hand, pointing , with umbrella and suitcase)		
Figure, plaster or chalkware, 9", "Mexico" molded on base in back	15.00	25.00

Diecast E.T., holds up finger, in back, and dressed in girl clothes, front, each-$20, MIP.

Cup topper-$10, MIP.

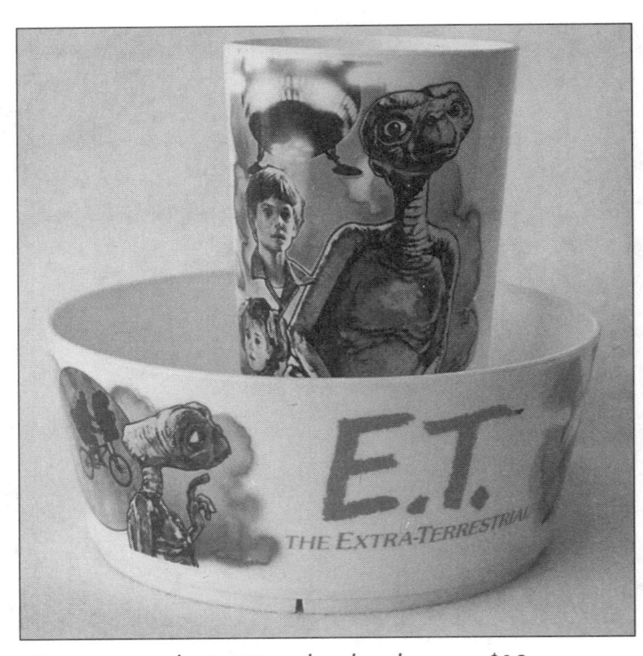

Dinnerware, plastic 20-oz. bowl and cup set-$18.

Dinnerware, plastic plate-$8.

Jumbo eraser-$5.

Drinking Glass, any design of four-glass set, 6 in., Pizza Hut, 1982, each-$9.

Item	Good/Loose	Mint/MIP
Figure, plaster, 16-1/2"	30.00	50.00
Finger, E.T. rubber finger, 8", (does not light up), "Hong Kong"	8.00	12.00
Finger Light, wearable, glows when pressed, 5", Knickerbocker, 1982	12.00	20.00
Game, E.T. the Extra-Terrestrial, Board Game, Parker Brothers, 1982	12.00	20.00
Game, E.T., Card Game, 6" x 4-1/4" box, Parker Brothers, 1982	10.00	20.00
Gift Set A, pack of six 2-1/2", figures, LJN, 1982	35.00	50.00
Gift Set B, pack of six 2-1/2", figures, LJN, 1982	35.00	50.00
Gift Set, Family: E.T., Elliott, Gertie, Michael, Mom, Bad Guy, LJN, 1982	40.00	60.00
Inflatable E.T., E.T. Pal, 3 ft. tall (life size), boxed, Coleco, 1982	35.00	50.00
Jewelry, assorted pieces on cards, Star Power, 1982, each	3.00	5.00
Key chain, round, tan, with E.T. portrait, 3" across, gold chain, 1982	8.00	12.00
Key chain, clear plastic with moon graphic, Greetings Unlimited, 1982	5.00	9.00
Key chain, 22K silver-plated, "He's Back - Extra Terrific," carded	8.00	10.00
Lamp, figural, ceramic, 13", light shows through heart	20.00	35.00
Life-size E.T. replica, Sharper Image, 1990s	1,000.00	1,300.00
Lobby Card set, 8 cards, 11" x 14", none show E.T., 1982, full set	25.00	40.00
Lunch Box with thermos, Aladdin, 1982	30.00	50.00

E.T. and Elliott Powered Bicycle-$20.

Item	Good/Loose	Mint/MIP
Novel, paperback, E.T. by William Kotzwinkle, 1982	3.00	6.00
Novel, hardback, E.T. by William Kotzwinkle, 1982	7.00	10.00
Paper cups, package, 1982	10.00	15.00
People magazine, cover story, book excerpt, June 28, 1982	5.00	8.00
People magazine, cover story, "All About E.T.," August 23, 1982	7.00	12.00
Pepsi box with E.T. video offer, 1988	4.00	8.00
Photo button, any design, originally sold on card, Star Power, 1982, each	2.00	4.00
Picture disk LP, sides show ET portrait/ bike against moon	12.00	20.00
Plush toy, large size, 16", brown with blue eyes, Kamar, 1982	15.00	30.00
Plush toy, leatherette, Kamar, 1982	15.00	25.00
Plush toy, leatherette with bean bag hands and feet, 14", Kamar, 1982	20.00	30.00
Plush toy, pink with hand raised, Applause, 1988	10.00	15.00
Plush toy, pink with arms folded across chest, 10", Applause, 1988	8.00	12.00
Plush toy, pink with vinyl head, green shirt, 7-1/2", U.C.S., 1996	5.00	10.00
Plush toy, brown with blue eyes, red smile, 8", Showtime, 1982	8.00	12.00
Plush toy, brown with blue eyes, 12", Showtime, 1982	10.00	16.00
Plush toy, unlicensed knock-off (several made), 1980s, each	8.00	12.00
Postcard, Spielberg with E.T.	1.00	2.00
Poster, E.T. with Michael Jackson, 1982	15.00	20.00
Poster, Pepsi mail-order premium, 1988	5.00	10.00
Puppet, vinyl figural puppet with red chest, blue eyes 6", U.C.S.	12.00	18.00

E.T. and Spaceship Launcher-$25.

Item	Good/Loose	Mint/MIP
Makeit & Bakeit Stained Glass craft set, any design, 1982	8.00	12.00
Message Pad, "E.T. Phone Home!" 1982	7.00	10.00
Movie Poster, one-sheet, 1982	50.00	75.00
Mug, ceramic, with figural E.T. handle, 4-1/2", Avon, 1983	12.00	20.00
Napkins, 9-7/8" x 10", Reed Mnf., 1982, full pack	5.00	8.00
Night Light, figural ET with arms folded on stomach, blue eyes, Star Power, 1980s	12.00	20.00
Night Light, brown E.T. with glowing chest, 4", Star Power, 1980s	8.00	12.00

Pop-up Spaceship-$20.

Figure-$12.

Figure, plaster or chalkware-$25.

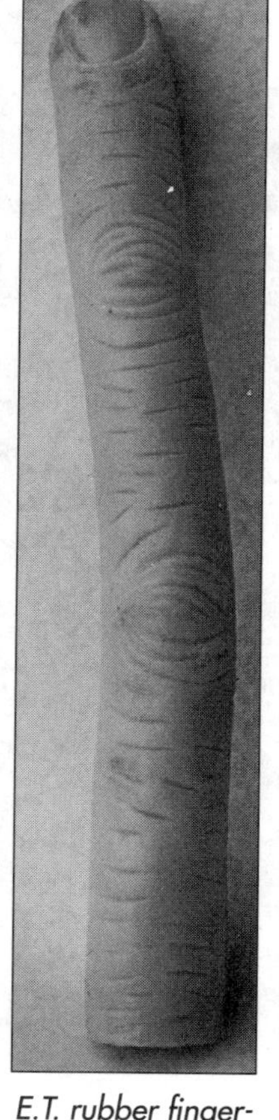

E.T. rubber finger-$12.

Item	Good/Loose	Mint/MIP
Puppet, E.T. Punching Puppet, brown face, polka-dot shirt, no mark, 1990s	10.00	15.00
Puzzle, frame tray, 15 pieces, various designs, Craftmaster, 1982	10.00	18.00
Record, "E.T. Speaks"	12.00	20.00
Record, soundtrack LP, boxed set with Michael Jackson narration, poster	30.00	50.00

Item	Good/Loose	Mint/MIP
Reese's Pieces box with walking E.T. figure, 1982	20.00	28.00
Riding Toy, E.T. head on front, child size, Coleco, 1982-83	50.00	75.00
Rolling Stone magazine, issue # 374, E.T. cover story, "A Star is Born," July 22, 1982	10.00	15.00
School Accessory Pack, Star Power, 1982	10.00	15.00
Scrapbook, with photos from movie, large trade paperback, 1982	7.00	10.00
Self-Adhesive Seals, 4 sheets per pack, various styles, Hallmark, 1982	3.00	6.00
Shrinky Dinks, by Colorforms, 1982	15.00	25.00
Soap, E.T. and Elliott decal soap, 3 oz., boxed, Avon, 1983	8.00	12.00
Sticker Album, with set of 120 stickers, Topps/Panini, 1982	22.00	28.00
Stickers, Reese's Collector stickers, 4-card set A, B or C, each	4.00	8.00
Stickers, puffy stickers, sheet of nine with card topper, 1982	5.00	10.00
T-shirt, E.T. hugs Elliott, with Reese's Pieces, 1982	12.00	20.00
T-Shirt, lights up, 1982	18.00	25.00
Talking E.T., plastic with pull string, 7" tall, boxed, LJN, 1982	20.00	35.00
Talking E.T., dressed in robe, 7" tall, boxed, LJN, 1983	25.00	40.00
Talking E.T. Telephone, 10", boxed, Hasbro Preschool, 1982	40.00	60.00
Tote Bag, Blue with E.T. graphic and tag, Star Power, 1982	12.00	20.00
Touch & Tell, electronic preschool game, 10" x 14", Texas Instruments, 1981	25.00	40.00
Trading Cards, Topps, 1982		
individual card	15 cents	25 cents
stickers #1-9, each	25 cents	50 cents
stickers #10-12, each	2.00	3.00
Full set (87 cards, 12 stickers)	25.00	30.00
unopened box	22.00	30.00
TV, Bed and Play Tray, 1982	12.00	20.00
View-Master gift set, E.T. and Elliott photo on box	15.00	25.00
Wallet, blue vinyl with E.T. moon graphic, U.C.S., 1982	20.00	30.00
Watch, Official E.T. watch, digital, shows bust of E.T., Nelsonic, 1982	40.00	60.00
Watch, Melody Glow Alarm, digital, shows E.T. full body, Nelsonic, 1982	40.00	65.00
Watch, clock style with hands, and E.T. on face, 1982	40.00	65.00
Wind-up Walker, glowing heart, LJN, 1982, carded	10.00	15.00
Wind-up Walker, dressed in scarf, robe or dress and hat, LJN, 1982	10.00	15.00
Window Grabber, plush with suction cup hands, "Phone Home" T-shirt, vinyl head, 9"	8.00	16.00

Board Game-$20.

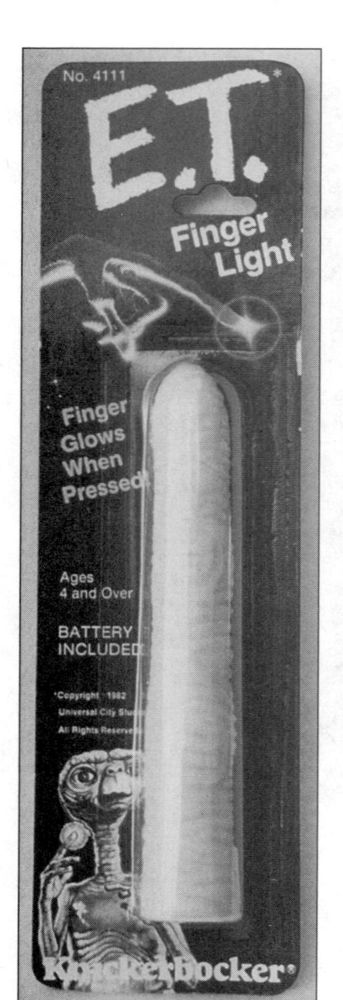

Finger Light-$20.

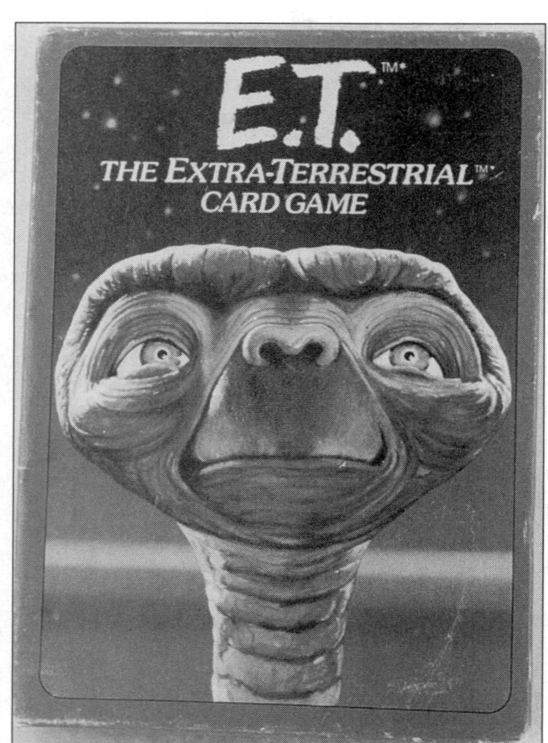

Card Game-$20.

Inflatable E.T.-$50.

Jewelry, assorted pieces on cards, each-$5.

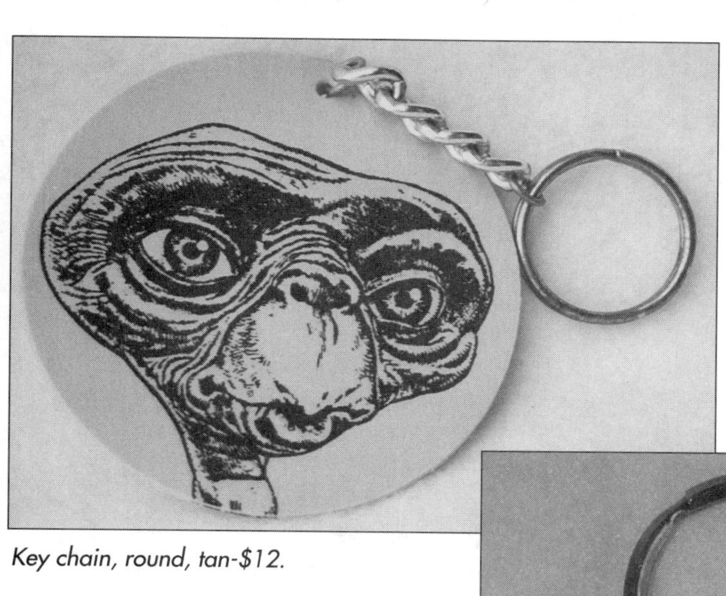

Key chain, round, tan-$12.

Key chain, 22K silver-plated-$10.

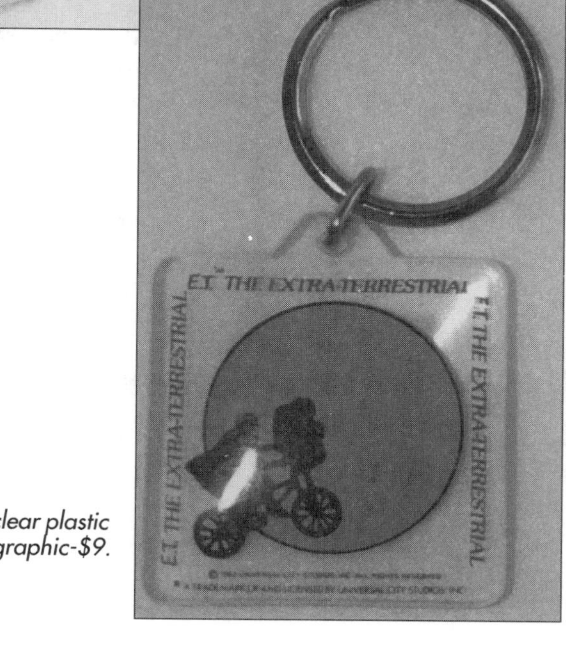

Key chain, clear plastic with moon graphic-$9.

Lunch Box with thermos, Aladdin, 1982-$50.

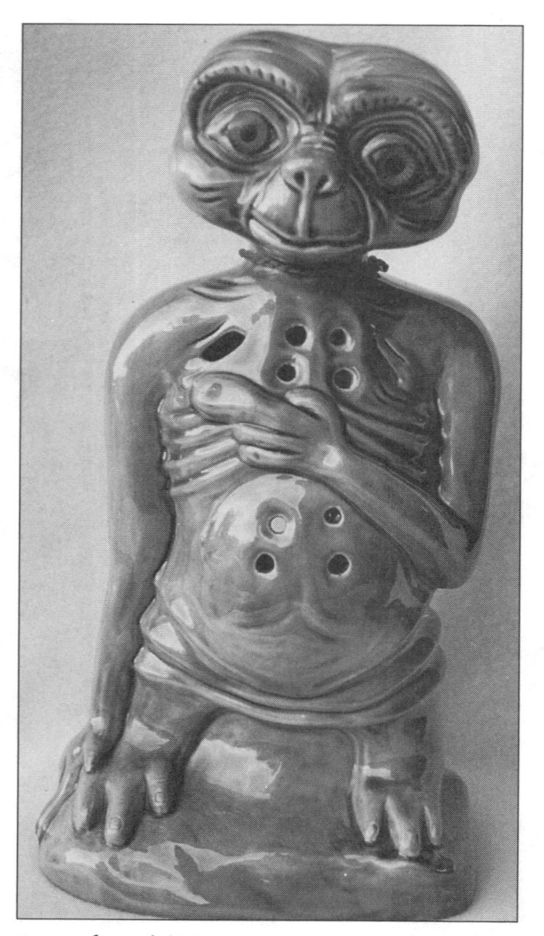

Lamp, figural-$35.

Mug, ceramic-$20.

Napkins-$8.

Night Light, brown E.T. with glowing chest-$12.

Night Light, figural ET with arms folded on stomach-$20.

People magazine-$12.

Picture disk LP-$20.

Plush toy, large size-$30.

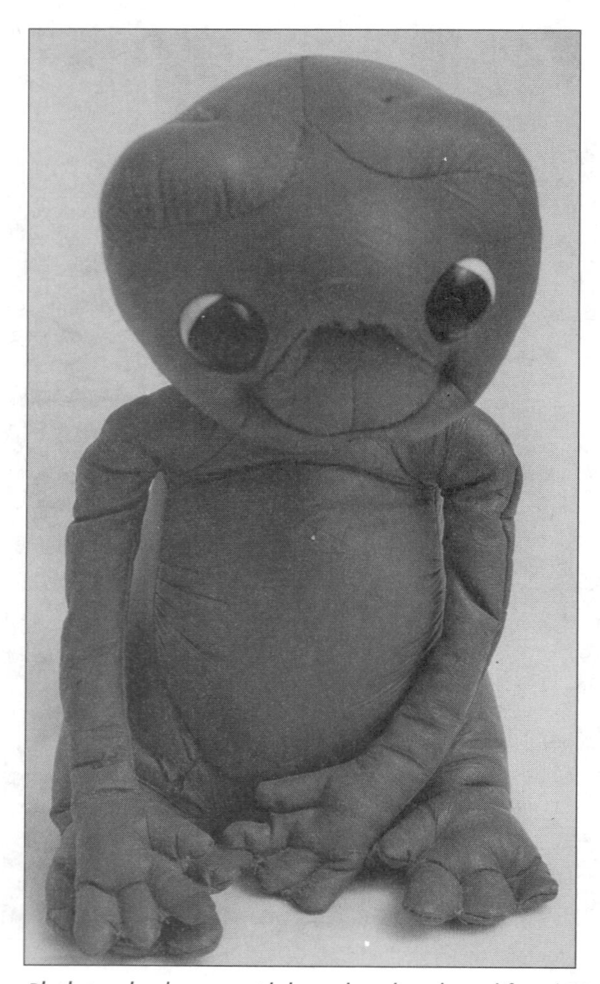

Plush toy, leatherette with bean bag hands and feet-$30.

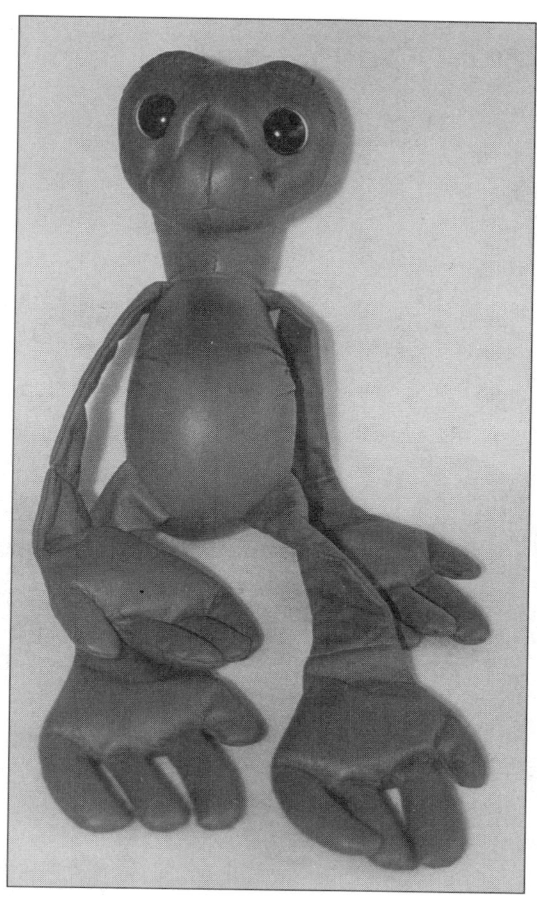

Plush toy, leatherette with bean bag hands and feet-$30.

Plush toy, pink with arms folded across chest-$12.

Plush toy, pink with vinyl head-$10.

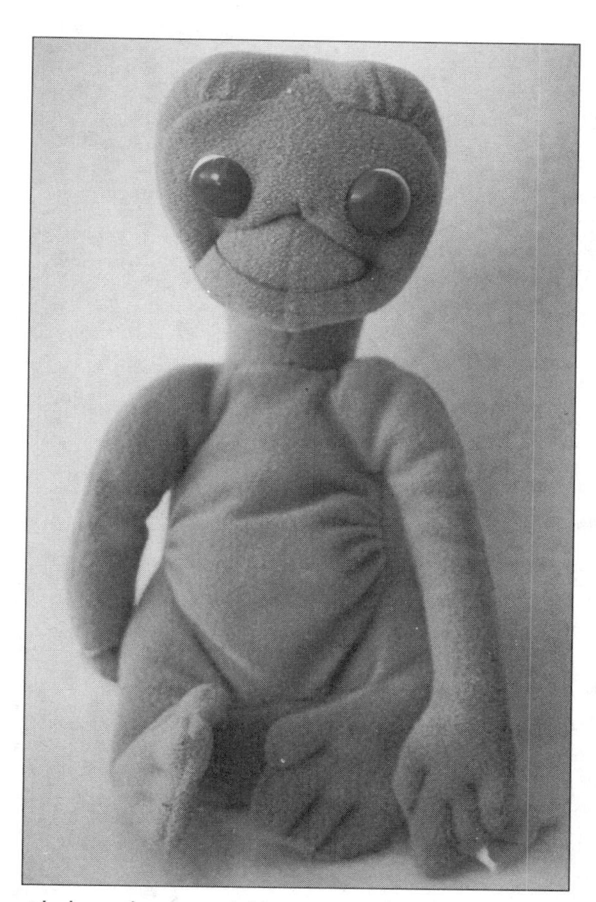

Plush toy, brown with blue eyes, red smile-$12.

Puzzles, frame tray, 15 pieces, various designs, Craftmaster, 1982-$18.

Punching Puppet-$15.

Record, soundtrack LP-$50.

Riding Toy-$75.

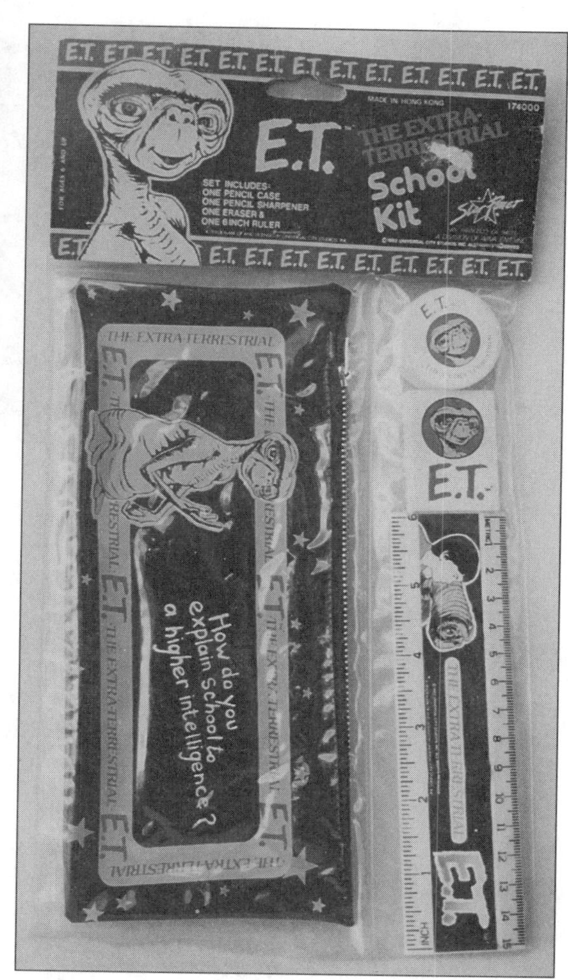

School Accessory Pack-$15.

Scrapbook, with photos from movie-$10.

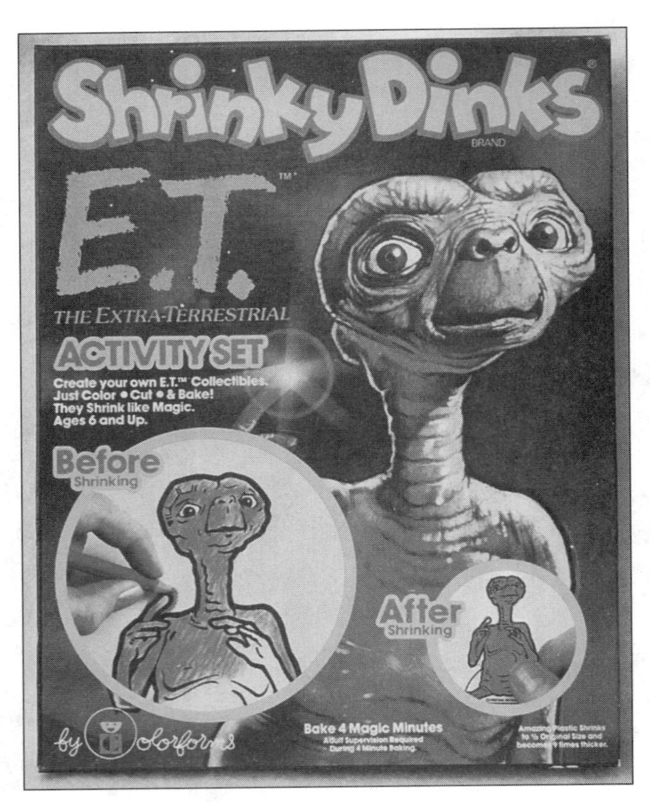

Shrinky Dinks-$25.

Soap-$12.

Talking E.T., dressed in robe-$40.

Talking E.T. Telephone-$60.

Tote Bag-$20.

Trading Cards, Topps, 1982.

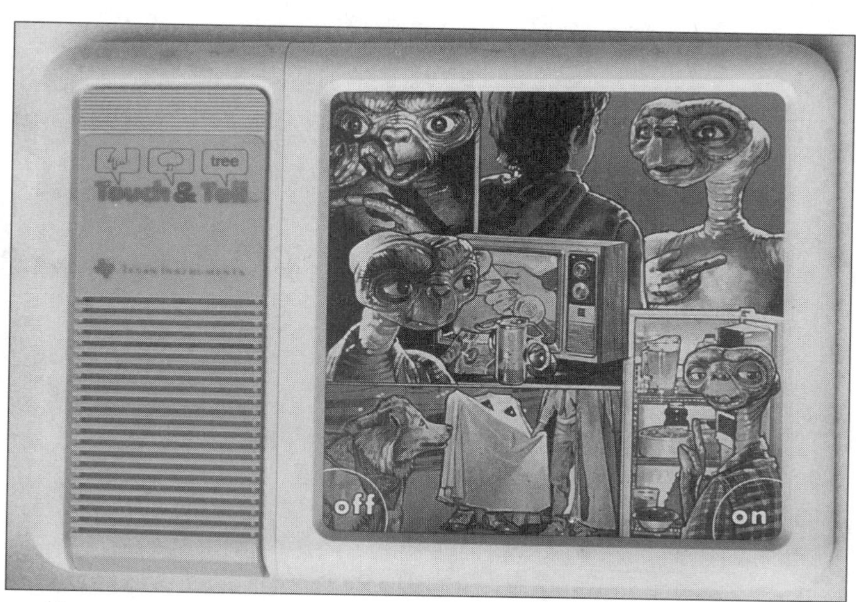

Touch & Tell, electronic preschool game-$40.

Wallet-$30.

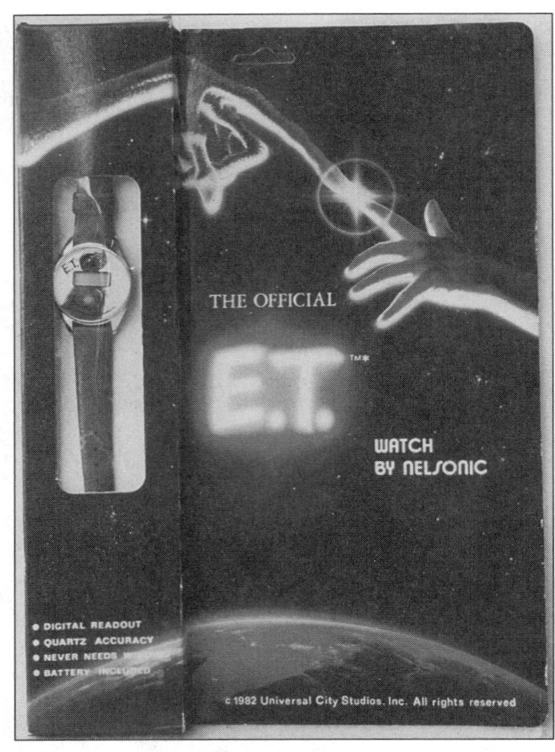

Official E.T. watch-$60.

Wind-up Walker, glowing heart-$15.

Window Grabber-$16.

Buck Rogers Flying Saucer, printed paper plates with metal rim, 6" across, S.P. Co.-$600.

Mystery Space Ship, 50+ tricks, 7" x 11" box, 8" diameter, Marx, 1960-$150.

Scorpio, purple alien, battery-op bendee from Major Matt Mason line, 8", Mattel, 1969-$350.

Waxy plastic figure, Milky Way, orange with dish head, 2", Miller, 1950s-$175.

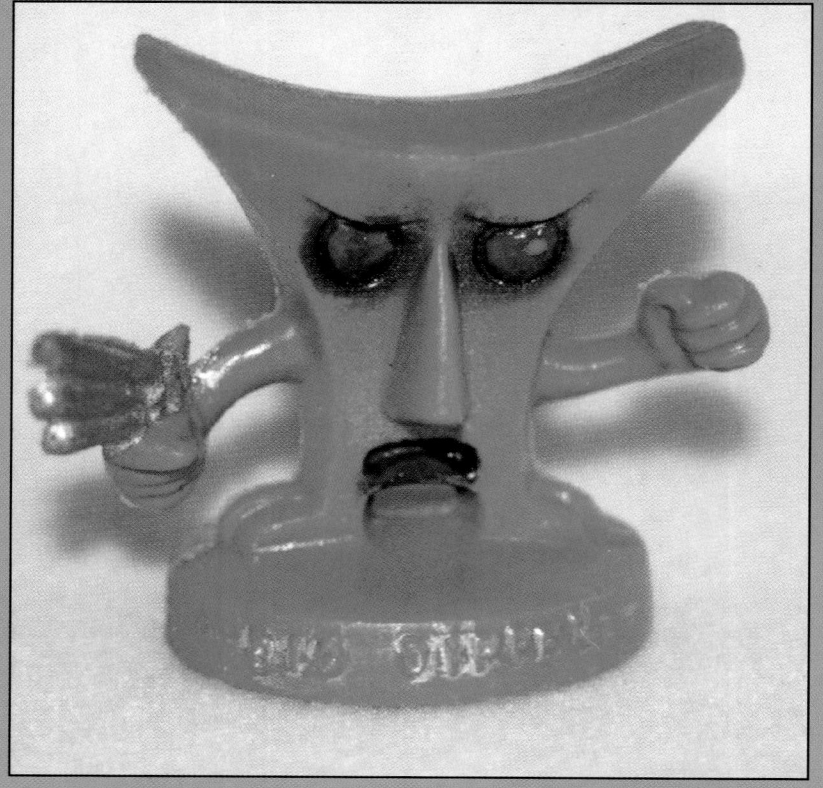

Waxy plastic figure, Big Dipper, green, 2", Miller, 1950s-$175.

Waxy plastic figure, Saturn, bird-like with gun, 4", Miller, 1950s-$175.

Waxy plastic figure, Purple People Eater, with silver trim, 4-3/4", Miller, 1950s-$175.

Waxy plastic figure, Pluto, dragon-like with blazing guns, 4" tall, Miller, 1950s-$175.

Fate, **published by Clark Publishing Co. 1948-1988, Llewellyn Publications, Vol. 1, #2, "Crow River Flying Disk?" Summer 1948-$30.**

Flying Saucers Are Real, **The, Donald Keyhoe, 1st printing, Fawcett Gold Medal, 1950-$30.**

UFO, **Robert Chapman, softcover, Mayflower Books (British), 1974-$10.**

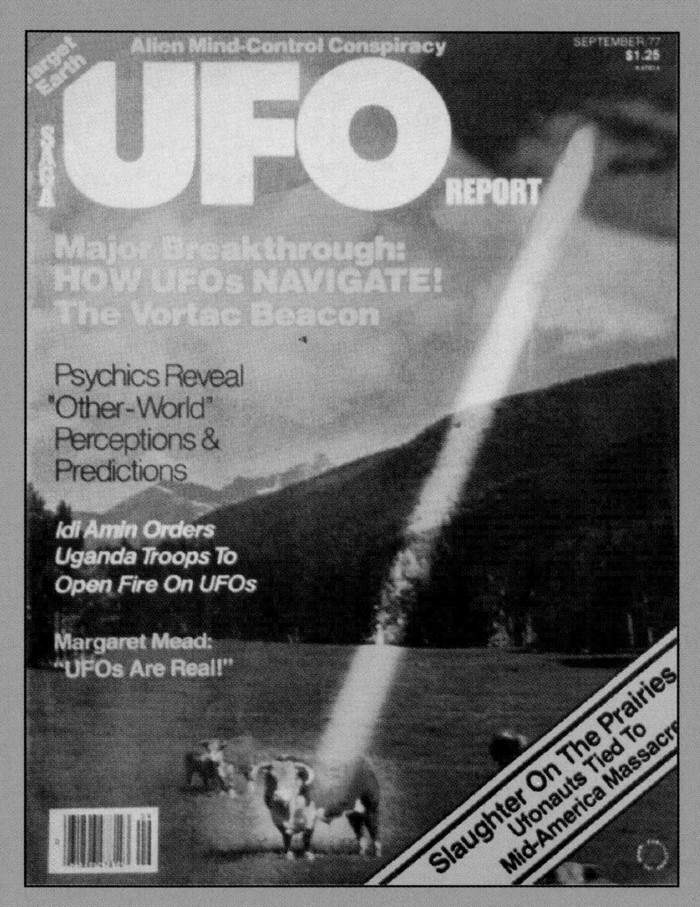

UFO Report **(Saga's UFO Report), published by Gambi Productions, Vol. 4, #5, "Slaughter on the Prairies," September 1977-$12.**

:omic book, *UFO Flying Saucers,* **#2, November 970-$12.**

UFOs 1969 (Flying Saucers and), **published by K.M.R. Publications, #3-$20.**

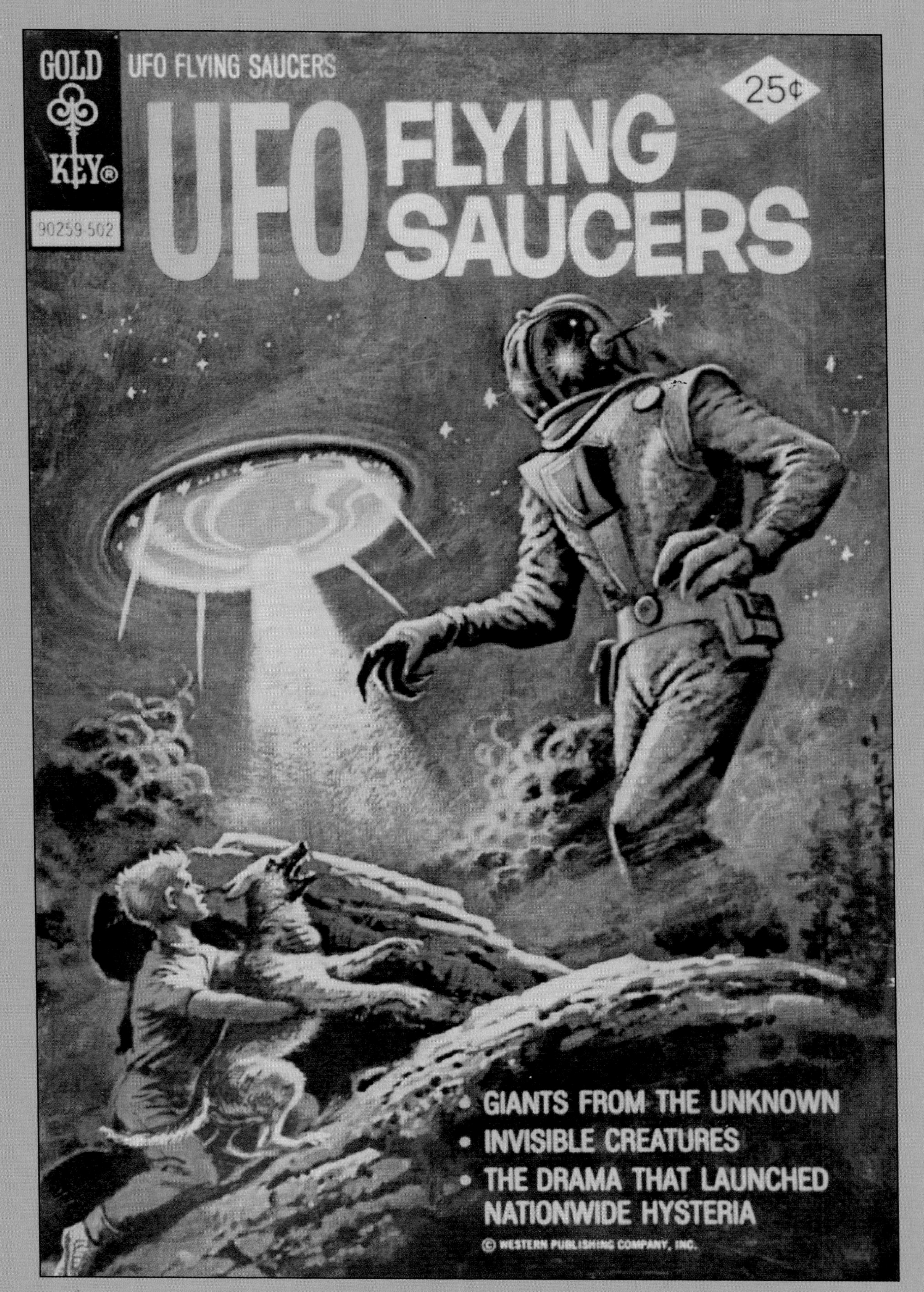

Comic book, *UFO Flying Saucers,* **#5, February 1975 - $10.**

Rolling Stone **magazine, issue #374, E.T. cover story, "A Star is Born," July 22, 1982-$15.**

E.T. figure, plaster, 16-1/2"-$50.

Marvin the Martian plate, Hare-Way to the Stars, Warner Bros., 1993-$20.

Rolling Stone, cover story on The X-Files (David Duchovny and Gillian Anderson in bed), U.S. version, May 16, 1996-$8.

Action figure from the movie *Mars Attacks,* Supreme Martian Ambassador, 12", boxed, Trendmasters, 1996-$20.

Movie poster for *UFO,* United Artists, 1956-$60.

Movie poster, Australian daybill, for the 1956 film, *Earth vs. The Flying Saucers*-**$300.**

Highway 285 runs through the heart of Roswell, N.M. If there is any doubt that you are in the UFO capitol of the world, it is quickly soothed as the restaurant and road signs become clear...

No collector should leave Roswell without hitting all of the major gift shops, including The Alien Zone, Starchild, Southwest Alien, Millennium Zone, Alien Spacecraft? UFO Research Center and the UFO Museum gift shop!

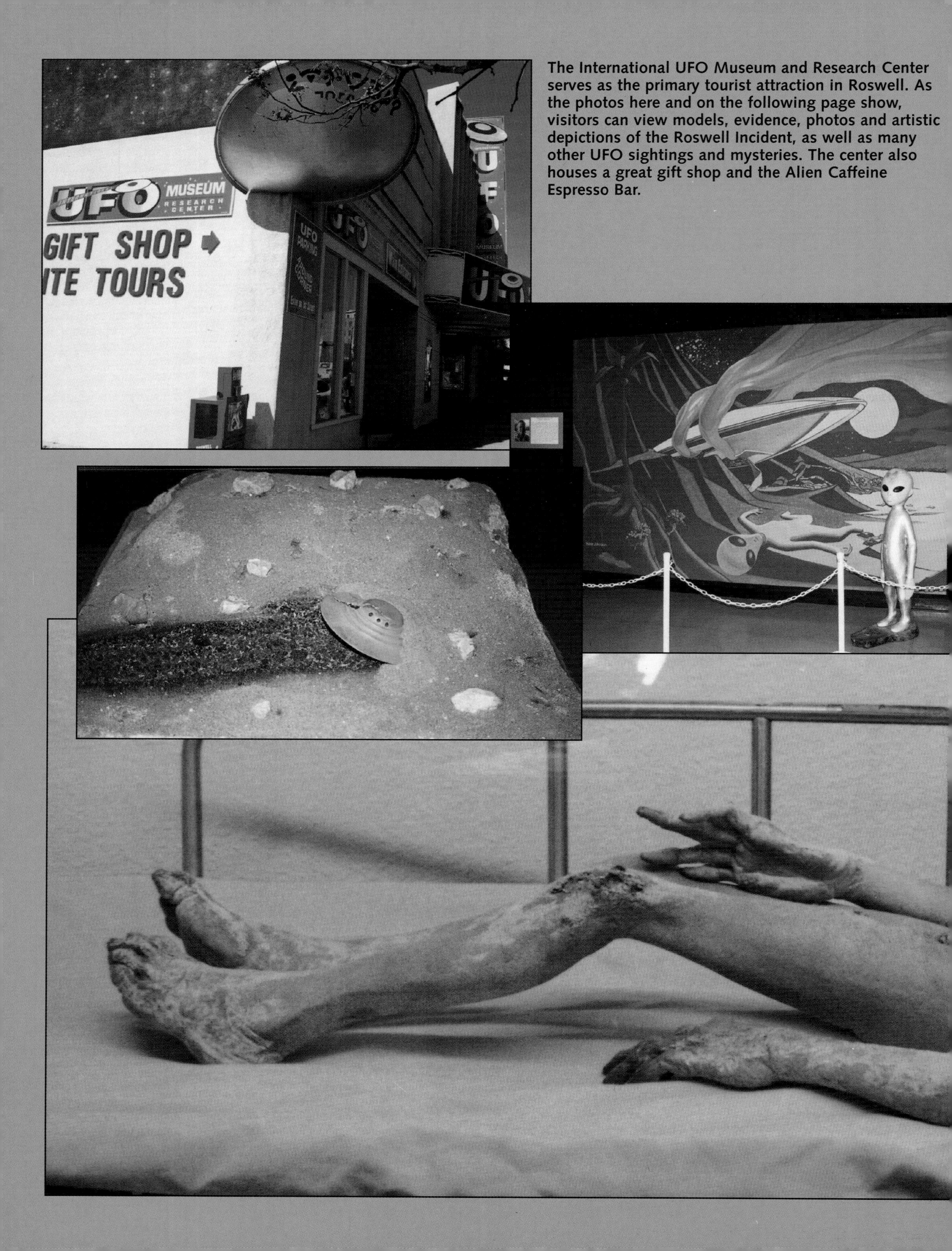

The International UFO Museum and Research Center serves as the primary tourist attraction in Roswell. As the photos here and on the following page show, visitors can view models, evidence, photos and artistic depictions of the Roswell Incident, as well as many other UFO sightings and mysteries. The center also houses a great gift shop and the Alien Caffeine Espresso Bar.

Alien
Caffeine
Espresso Bar
©1997

"Welcome Earthlings"
Del Norte Elementary School
Please do not touch displays

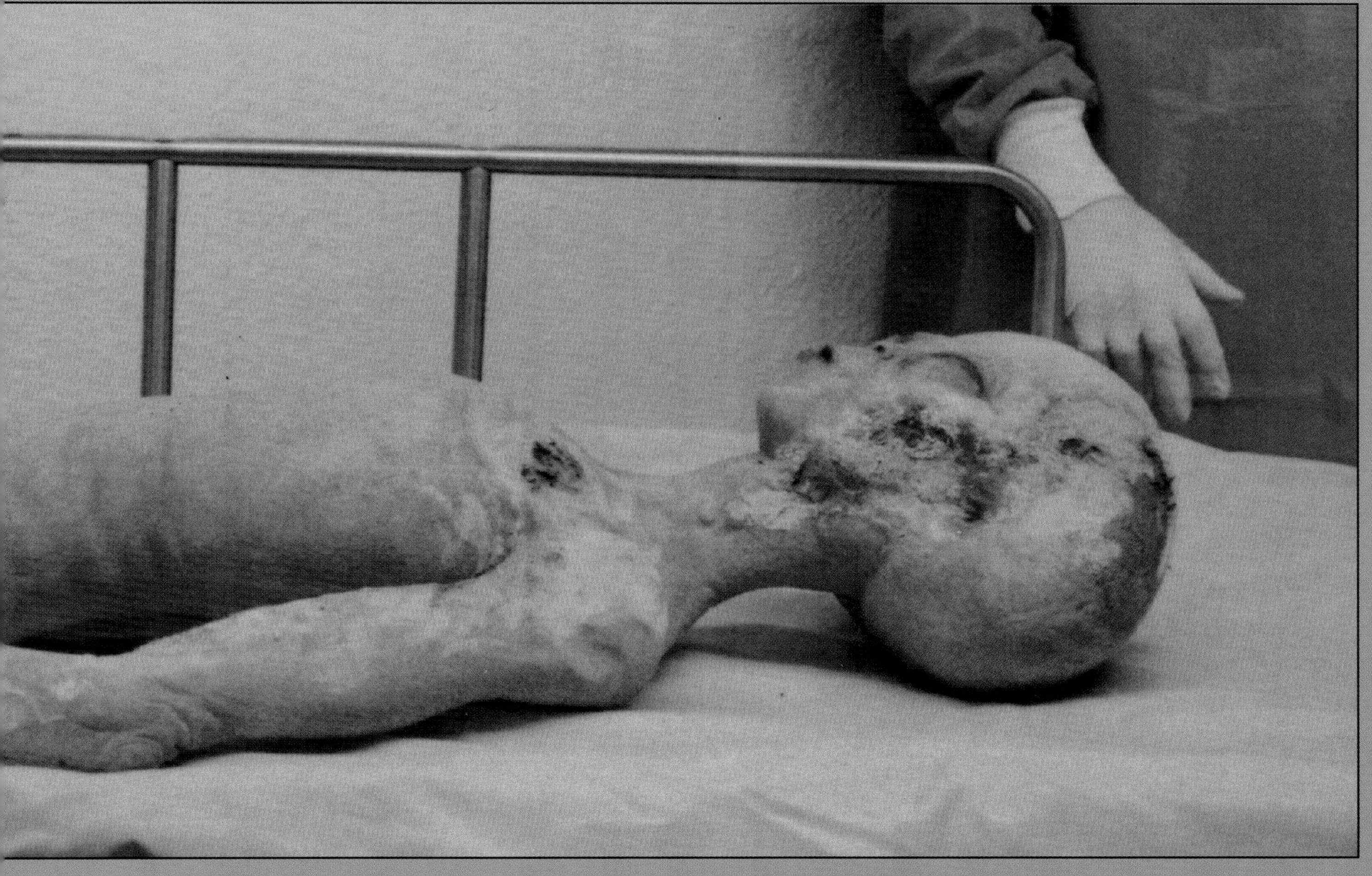

Area 51—"Roswell's 'Secret' Research Center" is one of Roswell's newest attractions, playing on the popularity and mystique of the "other" UFO hot spot in the southwest, Nevada's infamous Area 51. Visitors enter through a secret passageway in The Alien Zone gift shop and enjoy a photo opportunity paradise with a variety of comical life-size alien research dioramas, shown here and on the following page.

DEXTERITY ANALYSIS
FITNESS ANALYSIS
RELAXATION ANALYSIS
UFO
Walton

Remember children, never take candy from strangers...especially if they look like this!

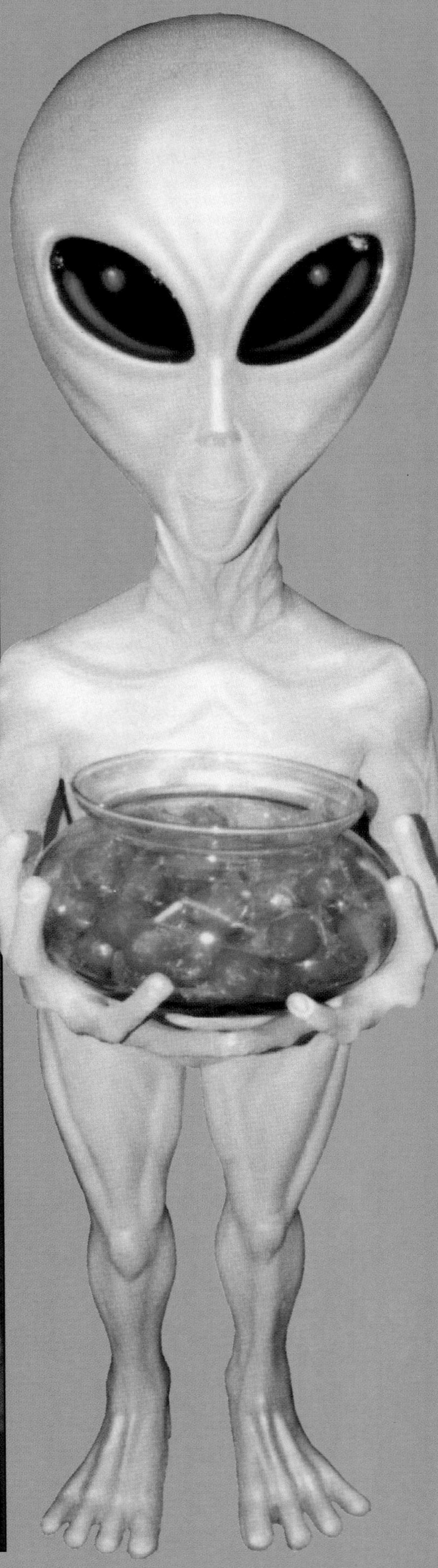

Local shop owners and employees in Roswell, such as this clerk from The Alien Zone, are always happy to talk about what's hot in spacey merchandise.

Chapter 4
It Came from the TV Set

Alf (1986-1990)

Known as Gordon Shumway on his home planet Melmac, Alf crash landed on earth's TV landscape in September 1986. This alien was like no other that had graced the airwaves. Short, sporting golden-brown fur, a big nose and an appetite for pet cats, Alf shocked his adoptive suburban family with his often twisted sense of humor. The wise-cracking alien had much to say about suburban earth culture, and the Tanner family spent much of its time trying to keep Alf hidden and out of trouble. Stand-out episodes include Alf battling a gigantic alien cockroach, Alf sighting Elvis and Alf becoming addicted to cotton, a mind-altering drug on his home planet.

Costume, Collegeville-$20.

The show spawned two Saturday morning shows, *Alf* and *Alf Tales*, along with a made-for-TV movie in which Alf is captured by the U.S. government.

Suction window grabber-$8.

Talking Alf plush toy-$25.

Item	Good/Loose	Mint/MIP
Alf Tales figures, set of 6: Alf Hood, Sleeping Alf, Little Red Riding Alf, Alf of Arabia, Three Little Pigs, Sir Gordon of Melmac, Wendy's premiums, 1990, each	2.00	5.00
Clip-on toy, 1980s	3.00	6.00
Comic books, *Alf*, Marvel Comics, March 1988-February 1992		
#1, photo cover, March 1988	2.00	4.00
#2-28	1.00	2.00
#29, with 3-D cover and 3-D glasses	2.00	4.00
#30-49	1.00	2.00
#50, photo cover, 52 pages, final issue, February 1992	2.00	3.00
Alf Annual #1-#3	1.00	2.00
Comics Digest #1, 1988, reprints Alf issues #1 and #2	1.00	2.00
Holiday Special #1, 1988, 68 pages	1.00	2.00
Holiday Special #2, 1989, 68 pages	2.00	3.00
Spring Special #1, 1989, 68 pages	1.00	2.00

Item	Good/Loose	Mint/MIP
Costume, Collegeville, 8-1/2" x 11" box, 1986	12.00	20.00
Door knob card, Burger King, 1987	1.00	3.00
Joke and riddle disc, Burger King, 1987	1.00	3.00
Lunch box, red plastic with decal, no thermos, 1980s	20.00	30.00
Phone, Alf phone, plush Alf attached to plastic phone, boxed, 1980s	45.00	75.00
Plush toy, 18", Coleco, 1986	12.00	20.00
Puppet with record set, Alf - Born to Rock, Burger King, 1987	5.00	8.00
Puppet with record set, Cooking with Alf, Burger King, 1987	5.00	8.00
Puppet with record set, Sporting with Alf, Burger King, 1987	5.00	8.00
Puppet with record set, Surfing with Alf, Burger King, 1987	5.00	8.00
Refrigerator magnet, Burger King, 1987	2.00	4.00
Sand mole, Burger King, 1987	2.00	4.00
Stickers, Alf Album Stickers, full set with album, Diamond, 1988	25.00	35.00
Suction window grabber, plush, 7-1/2", Coleco, 1988	4.00	8.00
Talking Alf plush toy, 18", Coleco/Alien Productions, 1986	15.00	25.00
Trading cards, Topps, (Series 1, 1987 or Series 2, 1988)		
individual card	10 cents	25 cents
sticker	25 cents	50 cents
full set with stickers	12.00	16.00

Alien Nation (1989-1990)

One of the earliest Fox network shows to garner any attention, *Alien Nation* lasted only one regular season before being canceled, reportedly because it was so expensive to produce. The series, based on a 1988 film starring James Caan, also spawned two made-for-TV movies in the 1990s.

The series was set in contemporary California, with the addition of shiploads and shiploads of fugitive aliens living in society and attempting to integrate peacefully. The show's central alien family was headed by George, a policeman, and his wife, Susan. (The aliens had all been issued pronounceable human names.) George's partner on the force, Matthew Sikes, was a human who became romantically involved with an alien, Cathy. Near the end of the season, Matthew found out that the aliens required three people to have sex. He also found out that his partner, George, was pregnant, and in one of the series' most memorable episodes, George gave birth with Matthew as mid-wife. Good stuff.

Item	Good/Loose	Mint/MIP
Banner, felt, triangular, black with logo, promotional item 1980s	10.00	18.00
Comic book, *Alien Nation*, movie adaptation, 68 pages, DC Comics, December 1988	2.00	3.00
Trading Cards, FTCC , 1990		
individual card	10 cents	25 cents
full set (60 cards)	12.00	15.00
unopened box	20.00	25.00

Captain Video (1949-1955)

The very first TV space show, *Captain Video* originally aired on the Dumont Network in June 1949. It told the story of Captain Video and the Video Rangers, a group which defended the earth against local and extraterrestrial threats in the future. Captain Video was an inventive genius, who designed a plethora of devices which aided the team in their battles. The show's initial run ended in 1955, but the series also prompted *The Secret Files of Captain Video* (which debuted in 1953) and *Captain Video's Cartoons* (1956).

Alien figures, Post Raisin Bran premium, hose on nose, above, and bird-like alien, each-$100.

Item	Good/Loose	Mint/MIP
Alien figure, Post Raisin Bran premium, blue hard plastic (hose on nose), 2-1/2", 1950s	60.00	100.00
Alien figure, Post Raisin Bran premium, blue plastic bird-like alien, 2-1/2", 1950s	60.00	100.00
Flying Saucer Ring, 1950s	200.00	300.00
Interplanetary Space Men/Supersonic Space Ships playset, boxed, 1950s	150.00	200.00

The Invaders (1967-1968)

This mid-season replacement show ran for 43 episodes on ABC in the late 1960s. Roy Thinnes starred as architect David Vincent, who witnessed the landing of an alien spacecraft. Subsequently, while he tried to convince everyone the aliens had landed, the aliens were trying to kill him. The aliens were able to assume human form, but could sometimes be detected by their mutant little fingers. Their objective was to take over the earth before their home planet died.

Book-$15.

Model kit, Invaders UFO, 1/72 scale, plastic, Monogram , 1979-$65.

Item	Good/Loose	Mint/MIP
Book, *The Invaders—Dam of Death*, Whitman Authorized TV Edition, 1967	8.00	15.00
Comic book series, Gold Key, October 1967-October 1968		
#1, October 1967, photo cover	20.00	60.00
#2, photo cover	15.00	50.00
#3, photo cover	15.00	50.00
#4, October 1968, photo cover	15.00	50.00
Model kit, Invaders UFO, 1968, Aurora	80.00	100.00
Model kit, Invaders UFO, 1975, Aurora	60.00	75.00
Model kit, Invaders UFO, 1/72 scale, plastic, Monogram , 1979	50.00	65.00
Model kit, re-issue of above, Monogram	15.00	25.00
Video, pilot episode, "Beach Head," aired Jan. 10, 1967, bootleg	10.00	20.00

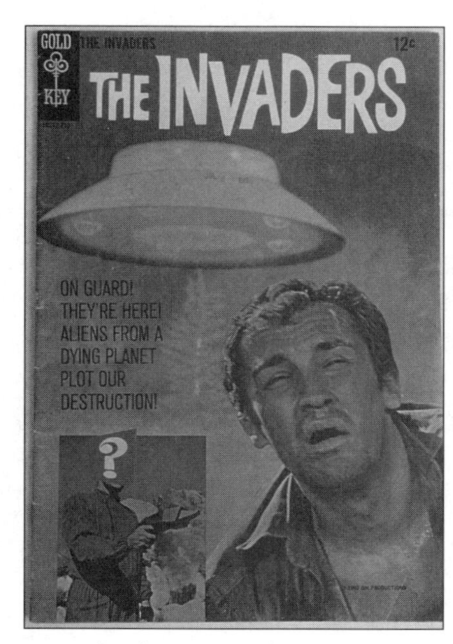

Comic book series, #1-$60.

Video, pilot episode-$20.

Marvin the Martian (1958-present)

While Marvin the Martian never had his own series, his popularity has soared after only a few appearances in Bugs Bunny cartoons. Marvin was introduced in 1958's "Hare-Way to the Stars," in which Bugs Bunny was rocketed to Mars and discovered Marvin (and his dog), plotting to blow up the Earth. More recently, he appeared with Michael Jordan in the feature film *Space Jam*. Today he is one of the most collected Warner Brothers' characters.

Item	Good/Loose	Mint/MIP
Alarm clock, Talking Alarm clock, 10", 1998	20.00	40.00
Bobble Head Figure, 7", resin, Warner Bros. Store exclusive, 1996	20.00	30.00
Figure, Marvin with dog in hinged rocket, 6" tall, resin, Warner Bros. Store exclusive, 1996	20.00	30.00
Figure, PVC, Marvin with flag, 2", 1988	5.00	10.00
Pez Dispenser, figural of Marvin and rocket, 6-1/2", carded, Pez, 1998	8.00	15.00
Plate, Hare-Way to the Stars, Warner Bros., 1993	12.00	20.00
Plush toy, 15", Ace, 1997	15.00	25.00
Slide puzzle, 3-D Slide puzzle, 4" x 4", DaMert Co., 1997	5.00	8.00
T-shirt, "Marvin the Martian," red with green and black ink, Warner Bros., 1993	15.00	20.00
Welcome mat, "Welcome to My Planet," rubber, 1990s	12.00	20.00

Talking Alarm clock-$40.

Figure, Marvin with dog in hinged rocket-$30.

Bobble Head Figure-$30.

Plush toy-$25.

Slide puzzle, 3-D-$8.

Figure, PVC-$10.

Welcome mat-$20.

Mork and Mindy (1978-1982)

Mork, the alien played by Robin Williams, first appeared as a character on *Happy Days*. In 1978, he was given his own series, *Mork and Mindy*, which became a big hit.

Mork arrived on Earth from his home planet Ork. He was sent here to study human behavior and earth culture, which interested those on his home world, where emotions were obsolete. At the end of each episode, Mork reported back to Orson, his leader on Ork, telling what he'd figured out about human emotions and life on earth. Mork lived with Mindy, a Boulder, Colorado, resident who decided to let him use her place as a home base to study our culture. The two fell in love over the course of the series, and in the final season, they got married and had "baby," played by Robin Williams' idol, Jonathan Winters.

Mork and Mindy Card Game-$25.

Mork and Mindy Game-$25.

Book-$20.

TV Guide-$8.

Item	Good/Loose	Mint/MIP
Action figure, Mindy, Mattel, 1970s	40.00	60.00
Action figure, Mork with egg, 4", Mattel, 1970s	35.00	50.00
Action figure, Mork with space pack, Mattel, 1970s	40.00	65.00
Book, *Video Novel*, with color photos, Pocket Books, 1979	10.00	20.00
Game, Mork and Mindy Card Game, with 4 Styrofoam eggs, Milton Bradley, 1978	15.00	25.00
Game, Mork and Mindy Game, Parker Brothers, 1979	15.00	25.00
Jeep, action figure accessory, Mattel, 1970s	20.00	35.00
Lunch box, metal photos on both sides (incl. Robby the Robot), King-Seely Thermos, 1979	35.00	55.00
Suspenders, official Mork suspenders, 1970s	25.00	35.00
Trading Cards, Topps, 1978		
individual card	15 cents	25 cents
sticker	25 cents	50 cents
full set (99 cards, 22 stickers)	12.00	16.00
unopened box	20.00	25.00
TV Guide, Mork and Mindy with egg on cover, Oct. 28-Nov. 3, 1978	4.00	8.00

My Favorite Martian (1963-1966)

When a Martian spaceship lands in his backyard reporter Tim O'Hare (Bill Bixby) adopts the pilot as his "Uncle Martin" (Ray Walston). This Martian appears as a human with antennae, and the laughs ensue.

The show inspired a cartoon series in the early 1970s, *My Favorite Martians*, in which a trio of Martians becomes friends with two Earthlings.

Item	Good/Loose	Mint/MIP
Cap, felt with spring antennae (bells on the ends), front patch with logo, picture, 1963	65.00	95.00

em	Good/Loose	Mint/MIP
oloring book, *My Favorite Martian*, Whitman, 1964	25.00	35.00
oloring book, *Uncle Martin the Martian Cut-Out Coloring Book*, Golden Press, 1964	28.00	40.00
oloring set, with palette, paints, crayons, etc., 13" x 18" box, Standard Toykraft, 1963	75.00	125.00
omic Book Series, Gold Key, January 1964-October 1966		
#1, January 1964, photo cover	50.00	100.00
#2, July 1964	20.00	60.00
#3, photo cover	20.00	50.00
#4-8	12.00	30.00
#9, photo cover	15.00	40.00
rayon by Number & Stencil Set, 10" x 15" box, Standard Toykraft, 1963	60.00	100.00
ame, "My Favorite Martian Game," Transogram, 1963	45.00	70.00
agic kit, Martian Magic Tricks, 14" x 20" box, Gilbert, 1964-64	150.00	200.00

'he Outer Limits (1963-1965)

Week after week, *The Outer Limits* captivated udiences with a parade of imaginative aliens and monsers. From the immortal opening credits, audiences knew ey were in for a weird ride. Showtime revived the series the early 1990s, debuting with a riveting version of eorge R. R. Martin's story, "The Sand Kings." Now runing on Fox as well, the new *Outer Limits* is proving to live p to its name, with intelligent and well-played stories.

Virtually every *Outer Limits* collectible produced ocuses on the show's aliens and monsters.

em	Good/Loose	Mint/MIP
nnual, large hardcover book, British, early 1960s, with comic strips	30.00	45.00
oard game, focuses on finding aliens, Milton Bradley, 1964	150.00	250.00

Comic books, Dell, #3 to 10-$30.

Mask, full head latex, Zanti Commander-$150.

Model kit, Chromoite, resin-$100.

Item	Good/Loose	Mint/MIP
Comic books, Dell, 1964-69		
#1	20.00	70.00
#2	12.00	40.00
#3-10	10.00	30.00
#11-18	7.00	25.00
Costume, Monster Mask and illustration on green suit, Collegeville, 1964	150.00	300.00
Famous Monsters magazine, cover story, issue # 26, January 1964	100.00	150.00

Item	Good/Loose	Mint/MIP
Mask, full head latex, Zanti Commander	100.00	150.00
Mask, full head latex, Sixth Finger	100.00	150.00
Mask, full head latex, Bug-eyed man	75.00	100.00
Model kit, Alien Soldier, resin, 12", name plate, Dimensional Designs, 1991	75.00	90.00
Model kit, Andro, resin, 7", with name plate, Dimensional Designs, 1991	85.00	110.00
Model kit, Chromoite, resin, 10", name plate, Dimensional Designs, 1991	80.00	100.00
Model kit, Ebonite Guard, resin, 11-1/2", name plate, Dimensional Designs, 1991	80.00	100.00
Model kit, Invisible Enemy, Lunar Models, 1990s	40.00	65.00
Model kit, OBIT Creature, resin, 11", name plate, Dimensional Designs, 1991	75.00	95.00
Model kit, Sixth Toe, super-deformed, Mad Lab, resin, 1993	15.00	25.00

Model kit, Zanti Commander-$95.

Model kit, Sixth Toe-$25.

Model kit, Thetan, resin, with name plate-$100.

Item	Good/Loose	Mint/MI
Model kit, Thetan, resin, with name plate, Dimensional Designs, 1991	75.00	100.0
Model kit, Thetan, resin, hand on ground, no name plate	65.00	95.0
Model kit, Venusian, resin and feathers, 13", name plate, Dimensional Designs, 1991	50.00	75.0
Model kit, Zanti Misfits, Lunar Models, 1990s	40.00	65.0
Model kit, Zanti Misfits, off-white resin, 3", Au-ora, unmarked Aurora tribute kit	50.00	75.0
Model kit, Zanti Misfits, super-deformed, resin, Monster Shop, unmarked, 1"	40.00	75.0
Model kit, Zanti Commander, 3-1/2", resin, name plate, Dimensional Designs, 1991	75.00	95.0
Outer Limits: An Illustrated Review, fanzine, issue #1, 1977	50.00	100.0
Outer Limits: An Illustrated Review, fanzine, issue #2, 1977	35.00	70.0
Outer Limits Newsletter, fanzine, 1978-79, 4 issues, each	10.00	20.0
Promotional custom TV monitor, hands hold TV with '90s logo	250.00	400.0
Puzzles, series of 6 100-piece jigsaw puzzles, Milton Bradley, 1964, each	75.00	125.0
Refrigerator magnet, rectangular, '90s logo	5.00	8.0
Trading cards, Bubbles Inc./Topps 1964		
Note: An unauthorized reprint set has also been produced.		
individual card	5.00	8.0
full set	350.00	500.0
unopened pack	100.00	150.0
wrapper	75.00	100.0
reprint set	5.00	10.0
Trading Cards, DuoCards, 1997		
individual card	15 cents	25 cen
full set (81 cards)	7.00	10.0
chromium card	4.00	5.0
gold card	20.00	25.0
card album	15.00	25.0
unopened box	35.00	40.0

Quisp (1960s-70s)

Quisp cereal, introduced by Quaker Oats in 1960, featured a wonderful TV ad campaign, animated by the great Jay Ward, creator of Bullwinkle and Dudley Doo-Right. Quisp was a funny little pink alien who has become one of the best loved and most collected cereal characters in history.

Item	Good/Loose	Mint/MIP
Bank, mail-order figural bank, 1960s	125.00	175.00
Model kit, Quisp—Quazy Energy, by Monsters in Motion, 1990s	75.00	100.00
Ring, flasher ring, shows Quisp then Quake, cereal premium, 1960s	50.00	70.00
Ring, meteorite ring, cereal premium, 1960s	30.00	50.00
Ring, super-sonic whistle ring, cereal premium, 1960s	30.00	50.00
Ring, ray gun ring, cereal premium, 1960s	40.00	60.00
Ring, Quisp plastic figural ring, 1" tall, cereal premium, 1960s	50.00	70.00

Trading cards, Bubbles Inc./Topps 1964, each-$8.

UFO (1972)

This British series was created by Gerry Anderson, best known for his "super-marionation" shows of the early 1960s (*Fireball XL-5*, *Thunderbirds*, etc.). Anderson's other "live actors" shows included *Space: 1999* and the 1990s' *Space Precinct*.

UFO was set in the 1980s, when a secret government agency, S.H.A.D.O. (Supreme Headquarters, Alien Defense Organization) was attempting to monitor an increasing alien threat. The show, although canceled early, attracted a cult following, perhaps because of its stylish sets and haircuts.

Item	Good/Loose	Mint/MIP
Diecast vehicle, S.H.A.D.O. UFO Interceptor, 8" long with plastic missile, Dinky #351	70.00	120.00
Video, Vol. 1-7, Polygram Video ITC, each	15.00	30.00
View-Master 3-reel packet with booklet	45.00	75.00

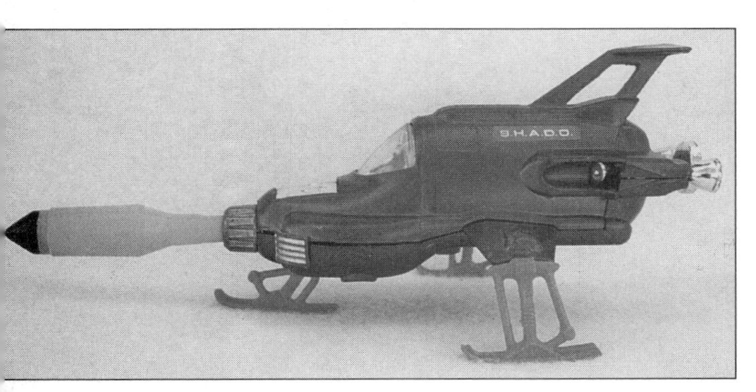

Diecast vehicle-$120.

View-Master, 3-reel packet-$75.

Video-$30.

V (1983-1985)

V was introduced to television as a mini-series in May 1983. A year later, a second mini-series, *V: The Final Battle*, was aired. That fall, the concept was given a regular series time slot on NBC.

V stood for "Visitors," the name given to invading aliens who posed as our friends. The Visitors were quickly exposed as the reptilian evildoers they were, and earth rebelled and took up the fight to oust them from our planet. But, it wasn't not easy.

Interesting relationships developed throughout the series, including inter-species relationships, half-breed children, and good-guy visitors. In the final episode of the series, a peace treaty was negotiated between humans and visitors. Marc Singer starred as the leader of the human resistance.

Item	Good/Loose	Mint/MIP
Book, *V*, paperback novel, Pinnacle, 1984	3.00	5.00
Books, various titles, paperback novels by Pinnacle, 1984-85, each	2.00	4.00
Bop bag, inflatable enemy alien, 48", boxed, 1980s	25.00	40.00
Comic book series, DC Comics, February 1985-July 1986		
#1, based on TV movie and series	2.00	3.00
#2-18, each	1.00	2.00
Doll, Enemy Visitor, removable human face reveals alien, tongue moves, 11-1/2", LJN, 1984	20.00	40.00
Gun, 45'er, Sound Pistol with holster, ricochet sound, carded, Arco, 1980s	30.00	40.00
Gun, 45'er Action Set, with pistol, grenade, binoculars, watch, carded, Arco, 1980s	35.00	50.00
Gun, M-16 Sound Rifle, machine gun sound, carded, Arco, 1980s	40.00	70.00
Lunch box with thermos, plastic with decal, Aladdin, 1984	30.00	45.00

Item	Good/Loose	Mint/M
Magazine, *Fangoria*, cover story, issue #39, 1980s	8.00	12.
Puzzles, jigsaw, 200-piece, 11" x 17", four designs, 1980s, each	8.00	15.
Trading cards, Fleer, 1984		
individual card	10 cents	20 ce
sticker	50 cents	1.
card set (66 cards)	8.00	12.
sticker set (22 stickers)	20.00	25.
unopened box	50.00	60.
Walkie Talkies, V Resistance, set of two, carded, Power Tronic, 1980s	40.00	65.

Doll, Enemy Visitor-$40.

Special Focus:
The X-Files

Since its 1993 debut on the Fox network, *The X-Files* has grown from a cult hit into a mainstream television icon. FBI special agents Fox Mulder and Dana Scully, played by David Duchovny and Gillian Anderson, have become familiar faces to all, appearing on numerous covers of magazines, *TV Guides* and books.

The X-Files series follows the adventures of Fox Mulder, an obsessive, alien-hunting, truth-seeking loner, and his FBI sidekick, Dana Scully, a logical scientist originally assigned to debunk Mulder's findings. The two became bonded by the end of season one, as Mulder began to respect Scully's "grounding" effects on his quest, and Scully began to realize there are things her scientific mind could not explain. The two spark an amazing on-screen chemistry, which has led to an entire subset of *X-Files* fans, known as "the Relationshippers," or, to those really in the swing of it, "the 'shippers."

The series owes its success to the chemistry of its two stars as much as it does to its wild storylines. Vacillating between anthology-style stories and soap-opera style continuing plot lines and conspiracies, the series mixes suspense, comedy and surrealism within its often-innovative scripts. Series creator Chris Carter is to thank for that.

Series highlights have included Mulder almost finding his sister, who was abducted by aliens, several times; Scully being abducted and impregnated by aliens; Mulder being captured by Russian scientists; Mulder and Scully both losing family members; and near fatal run-ins with a wide variety of monsters, mutants, extraterrestrial life forms, shape-shifters, ghosts, bugs, clones and more. The team has also uncovered an incredible number of government conspiracies, most of which seem to be intertwined in a gigantic galactic conspiracy involving various alien entities.

More than once, the FBI has closed the X-Files, and sent Mulder and Scully off to do more menial work, but it never seems to stop them. They manage to fight their way back and return, in the end, to business as usual—or in their case, business as UN-usual as it can possibly get.

The series inspired a feature film in 1998, *X-Files: Fight the Future,* which attempted to tie up a few loose ends, while not solving TOO much of the mystery.

Agent Dana Scully, with FBI clothes and alien on gurney-$15.

Agent Dana Scully, with FBI clothes and cryopod-$15.

Item	Good/Loose	Mint/MIP
Action Figures, McFarlane Toys, 1998		
Agent Dana Scully, 6", with FBI clothes and alien on gurney	10.00	15.00
Agent Dana Scully, 6", with FBI clothes and cryopod	10.00	15.00
Agent Fox Mulder, 6", with FBI clothes and alien on gurney	10.00	15.00
Agent Fox Mulder, 6", with FBI clothes and cryopod	10.00	15.00
Agent Mulder, 6", with Arctic gear and cryopod	10.00	15.00
Agent Mulder, 6", with Arctic gear and alien	10.00	15.00
Agent Scully, 6", with Arctic gear and cryopod	10.00	15.00
Agent Scully, 6", with Arctic gear and alien	10.00	15.00
Attack Alien, 6", with caveman	10.00	15.00
Fireman, 6", either color variation	12.00	20.00

Agent Fox Mulder, with FBI clothes and alien on gurney-$15.

Agent Mulder with Arctic gear and cryopod-$15.

Item	Good/Loose	Mint/MIP
Badge, FBI Special Agent, Scully or Mulder fake ID, 4" x 2-3/4" laminated badge with clip, 1990s	3.00	5.00
Book, Goblins, by Charles Grant, Harper Prism paperback, 1994	3.00	6.00
Book, Official Guide to the X-Files, Brian Lowry, 6-1/4" x 9-1/4" softcover, Harper, 1995	15.00	20.00
Book, Science of the X-Files - The Truth, by Michael White, Legend (British import), 1990s	8.00	12.00

Item	Good/Loose	Mint/MIP
Book, Scully X-Posed, unauthorized bio, Nadine Crenshaw, 8-1/2" x 11" softcover, Prima, 1997	15.00	25.00
Book, Trust No One: Official Third Season Guide, by Brian Lowry, Harper Prism	12.00	16.00
Book, Unauthorized X-Cyclopedia, Hatfield & Burt, 6" x 9" softcover, Kensington Books, 1977	12.00	16.00
Book, Unofficial X-Files Companion, by N.E. George, 7-1/2" x 9", softcover, 1990s	10.00	15.00

Agent Scully, with Arctic gear and cryopod-$15.

Attack Alien, with caveman-$15.

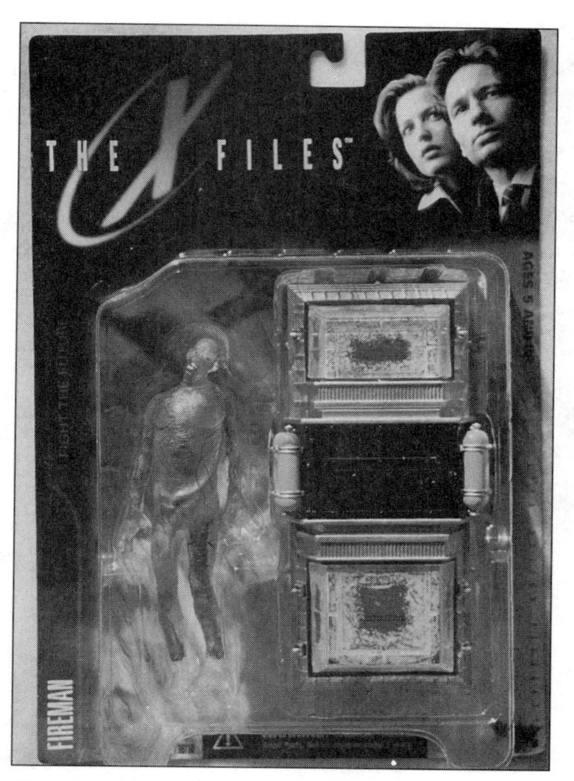

Fireman, either color variation-$20.

Badge, FBI Special Agent, Scully or Mulder fake ID-$5.

Book, The Truth is Out There, the Official Guide to The X-Files-$20.

Book, Science of the X-Files - The Truth, (British import)-$12.

Book, unauthorized bio-$25.

Item	Good/Loose	Mint/MIP
Bust, Mulder or Scully, 10", Legends, 1990s	150.00	175.00
Comic Book Series, Topps Comics, January 1995-present		
#1/2, with certificate, 1995	10.00	25.00
#1, direct market or newsstand edition	20.00	40.00
#2	10.00	30.00
#3	5.00	15.00
#4	5.00	15.00
#5-25, each	2.00	5.00
Annual #1	2.00	5.00
Hero Illustrated Giveaway	7.00	15.00

Item	Good/Loose	Mint/MI
Special Edition #1, reprints issues 1-3	4.00	8.0
Special Edition #2, reprints issues 4-6	2.00	5.0
Special Edition #3, reprints issues 7-9	2.00	5.0
Comic Book Trade Paperback, 6-1/2" x 10-1/2", Topps Comics	15.00	20.0
Comics Digest, quarterly beginning December 1995, Topps Comics, each	3.00	5.0
Computer Game, Fox Interactive game, box is 8-1/2" x 10", 1990s	25.00	35.0

Trust No One: Official Third Season Guide-$16.

Unauthorized X-Cyclopedia-$16.

Unofficial X-Files Companion-$15.

Comic Book Trade Paperback-$20.

Computer game-$35.

Key chain-$5.

Official X-Files Map-$30.

Item	Good/Loose	Mint/MIP
Dolls, X-Files Barbie & Ken, set #19630 with long-haired Scully, Mattel, 1998	75.00	125.00
Dolls, X-Files Barbie & Ken, set with normal-haired Scully, Mattel, 1998	65.00	80.00
Key chain, X-Files Light-Up Key chain, carded, 4-1/2", IPI, 1996	2.00	5.00

Pendant-$15.

Item	Good/Loose	Mint/MIP
Map, Official X-Files Map, shows where events occurred, Harper Collins, British import.	20.00	30.00
Model kit, X-Files: Paranormal Activities, 5th Sense, 1990s	125.00	175.00
Model kit, Flukeman, Dark Horse, 1990s	100.00	150.00
Model kit, Mulder and Scully, by Men in Black, 1990s	175.00	225.00
Mouse Pad, X-Files, various styles, each	10.00	15.00
Pendant, for necklace, "X" logo, pewter, 1990s	10.00	15.00

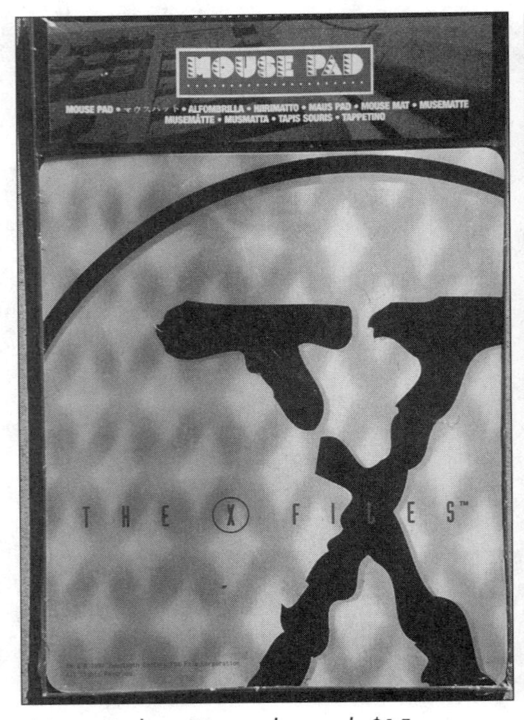

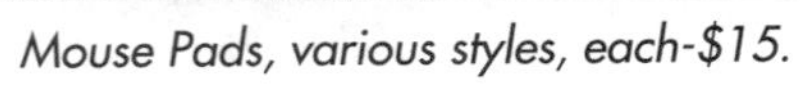
Mouse Pads, various styles, each-$15.

Item	Good/Loose	Mint/MIP
Postcard Book, "The Conspiracies," 7" x 4-1/2", Harper Prism, 1996	6.00	10.00
Poster, various styles, show Duchovny, Anderson with cryptic type, 1990s, each	6.00	12.00
T-shirt, Fight the Future, small type on front, big graphic on back, 1998	12.00	20.00
T-shirt, Karval Kon XX, rare, 1996	20.00	30.00
Trading Cards, X-Files, series 1, Topps, 1995		
set of 72 regular cards	8.00	12.00
set of 72 foil parallel version cards	40.00	50.00
etched foil card (6 different), each	6.00	8.00
Topps Finest card (4 different), each	6.00	9.00
card album	15.00	20.00
unopened box	50.00	65.00
Trading Cards, X-Files, series 2, Topps, 1996		
set of 72 regular cards	8.00	12.00
set of 72 foil parallel version cards	50.00	60.00
etched foil card (6 different), each	6.00	8.00
Hologram (4 different), each	8.00	10.00
unopened box	30.00	35.00
Trading Cards, X-Files, series 3, Topps, 1996		
set of 72 regular cards	8.00	12.00
set of 71 parallel version cards	50.00	60.00
etched foil card (6 different), each	4.00	5.00
Hologram (2 different), each	4.00	5.00
Paranormal Finest card (2 different), each	4.00	5.00
unopened box	30.00	35.00
Trading Cards, X-Files MasterVisions, Topps, 1996		
set of 30 cards	30.00	35.00
Trading Cards, X-Files Movie, Topps, 1998		
set of 72 regular cards	8.00	12.00
Mystery Card (6 different), each	5.00	8.00
autographed card, each	40.00	60.00
Trading Cards, X-Files Showcase, Topps, 1997		
set of 72 cards	10.00	15.00
X-effect card (6 different), each	5.00	7.00
laser card (6 different), each	6.00	8.00
unopened box	55.00	60.00
Trading Cards, X-Files Contact, Intrepid, 1997		
set of 90 cards	15.00	20.00
foil card (9 different), each	6.00	8.00
3-card foil strip (3 different), each	20.00	25.00
acetate card (3 different), each	8.00	10.00
3-card acetate strip	40.00	50.00
unopened box	50.00	60.00
Video, special collector's edition of *X-Files: Fight the Future* movie, with poster, booklet, script, 35mm film strip , etc., in 11-1/2" x 9-1/2" box	75.00	95.00

Poster, various styles, each-$12.

T-shirt, Fight the Future-$20.

Postcard Book-$10.

T-shirt, Karval Kon XX, rare-$30.

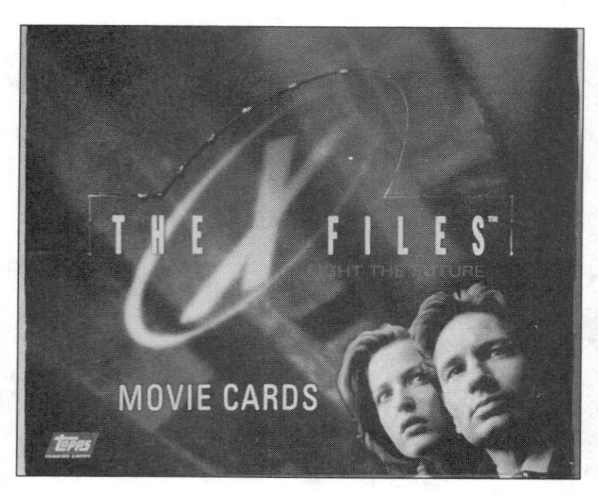

Trading Cards, X-Files Movie, Topps, 1998-set of 72 regular cards-$12.

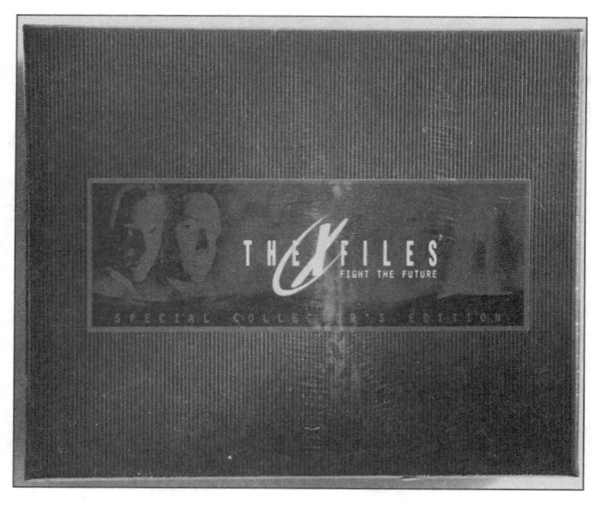

Video, specia collector's edi tion of X-Files Fight the Futu movie-$95.

Item	Good/Loose	Mint/MIP
Video promotion, vinyl poster, 1998	10.00	18.00
Watch, round case, ltd. ed., pewter finish, 10,000 made, Fossil, 1990s	50.00	75.00
Watch, round case, ltd. ed., sterling silver, 1,000 made, Fossil, 1990s	100.00	150.00

Magazine Covers

Item	Good/Loose	Mint/MIP
Cinefantastique, cover story, June 1998	7.00	10.00
Details, cover story, Anderson in silver make-up with alien hologram, June 1998	3.00	5.00
Entertainment Weekly, cover story, X-Files Movie Mysteries, July 10, 1998	3.00	5.00
Entertainment Weekly, part of cover montage, Oct. 16, 1998	3.00	5.00
Fate magazine, cover story, "Paranormal Invades Primetime TV," Nov. 1994	2.00	5.00
Newsweek, cover story, June 22, 1998	3.00	5.00

Item	Good/Loose	Mint/MIP
People Magazine, cover story (Duchovny, Anderson), October 9, 1995	8.00	12.00
Rolling Stone, cover story (Duchovny, Anderson in bed), US version, May 16, 1996	4.00	8.00
Rolling Stone, cover story (Duchovny, Anderson in bed), early Australian version, 1996	10.00	20.00
Rolling Stone, cover story, (Anderson with Creature from Black Lagoon), Feb. 20, 1887	6.00	12.00

Autographed card, each-$60.

Video promotion, vinyl poster-$18.

Item	Good/Loose	Mint/MIP
TV Guide, cover story, (Anderson, Duchovny look blue on cover), March 11-17, 1995	5.00	10.00
TV Guide, cover story, (Anderson, Duchovny embrace on cover), April 6-12, 1996	5.00	10.00
TV Guide, cover story (Anderson, Carter, Duchovny pictured), Nov. 15-21, 1997	4.00	8.00
TV Guide, cover story (two covers: Duchovny -X Man; Anderson-X Woman), June 20-26, 1998	3.00	6.00
US Magazine," David in Love" cover story (Duchovny), March, 1998	4.00	8.00
US Magazine, Anderson Licks Duchovny on cover, May 1997	6.00	10.00
Vanity Fair, cover story, "Lord of the Files," (Duchovny), June 1998	4.00	8.00
UFO Magazine, cover story, "The Real X Files," with movie photos, May/June 1998	4.00	8.00
Yahoo!, cover story, July 1998	7.00	10.00
X-Files Magazine, oversized, British import with free poster inside, November, 1998	12.00	20.00
X-Files Official Magazine, MVP Licensing Inc.		
#1, Spring 1997 (black and white cover photo)	10.00	15.00
#2, Summer, 1997	5.00	8.00
#3, Fall, 1997	4.00	7.00
#4, Winter, 1997-98	4.00	7.00
#5, Spring, 1998 (two covers available)	4.00	7.00
#6, Summer, 1998	4.00	7.00
#7, Fall, 1998	4.00	7.00
X-Files Movie - Official Magazine, Summer, 1998	4.00	7.00

Cinefantastique-$10.

Entertainment Weekly-$5.

Details-$5.

Newsweek-$5.

People magazine-$12.

Rolling Stone-$12.

TV Guide-$8.

TV Guide-$10.

TV Guide, (two covers: Duchovny -X Man; Anderson-X Woman), each-$6.

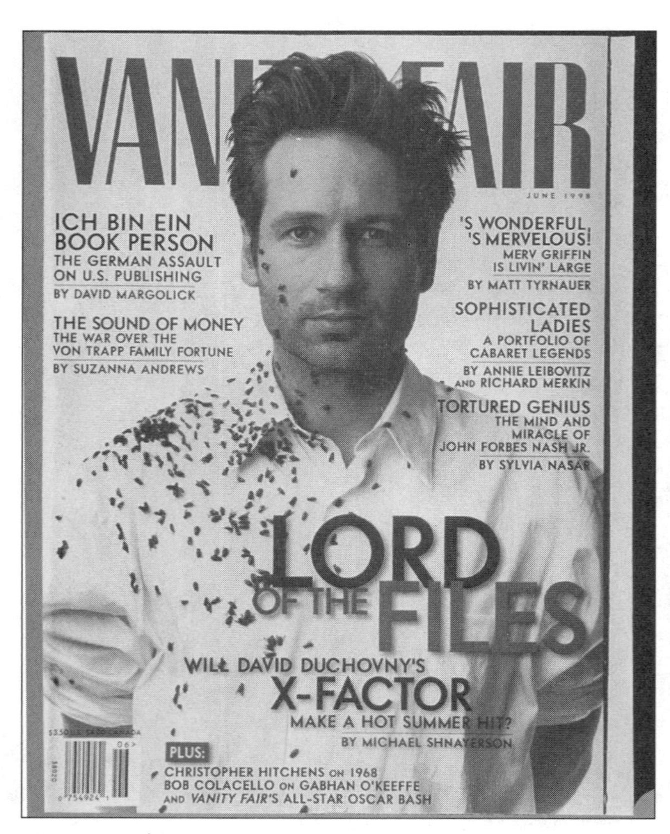

Vanity Fair-$8.

UFO magazine-$8.

US magazine-$8.

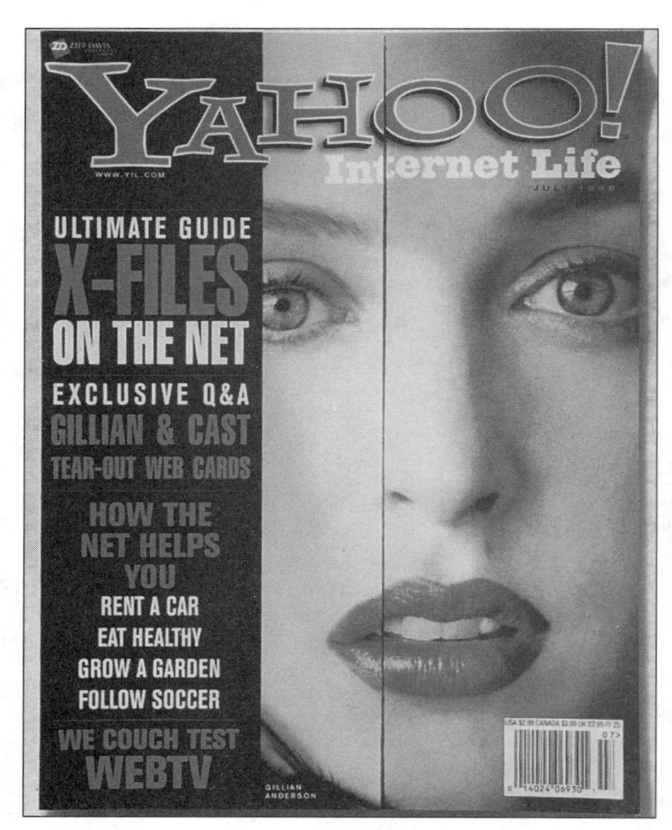

Yahoo!-$10.

X-Files Official Magazine, #1-$15.

X-Files Official Magazine, #2-$8.

X-Files Official Magazine, #3-$7.

X-Files Official Magazine, #7-$7.

X-Files Official Magazine, #6-$7.

X-Files Official Magazine, #4-$7.

X-Files magazine-$20.

X-Files Movie-Official Magazine-$7.

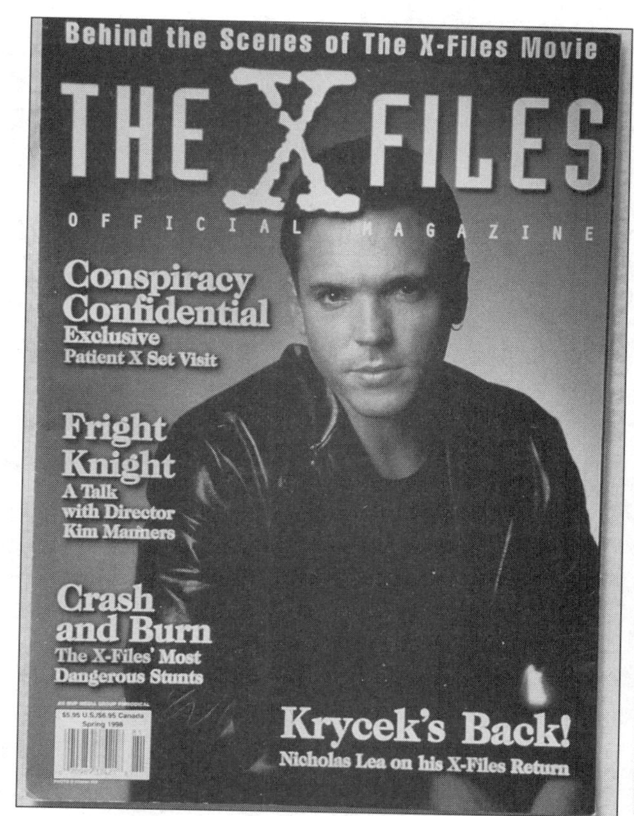

X-Files Official Magazine, #5 (two covers available)-$7.

Chapter 5
Aliens at Home

Aliens have become a household concept, and by the late 1990s, the classic, stylized image of a visiting "Grey" was one of the staples of hip fashion. Shopping mall gift shops were literally brimming with alien images, memorabilia, knick-knacks, T-shirts and more. Due largely to Fox TV's successful X-Files series, the infamous Alien Autopsy footage and the 50th anniversary of the Roswell Incident, the alien face graphic became a sort of modern day "smiley face."

Bead curtain-$25.

Candle-$12.

Apron-$25.

Item	Good/Loose	Mint/MIP
Apron, "UFO—Unidentified Frying Object," alien graphic on white background	15.00	25.00
Bead curtain, with alien head beads, boxed, Matscot, 1996	18.00	25.00
Candle, glow alien head candle, 4", no mark	7.00	12.00
Car antenna head, various colors, Styrofoam heads, Outtaspace Enterprises, 1990s	3.00	5.00
Christmas ornaments, aliens in various X-Mas situations, Shadow Box, 1998, each	6.00	10.00
Christmas stocking, fuzzy, with alien on top	15.00	20.00
Diary, "Space Log," with pen, 6-1/4" x 5-1/2", 1990s	5.00	8.00
Key chain, figural alien head, 1-1/2", no mark	1.00	3.00
Key chain, slinky alien, glows, 2-1/2"	3.00	6.00
Key chain, "Image Bottle," alien with glitter in vial of liquid, 2-1/2", 1990s	2.00	4.00
Lamp, figural alien head base, Matscot Intl. Inc., 1997	25.00	35.00

Car antenna head-$5.

Item	Good/Loose	Mint/MIP
Mug, ceramic, figural alien head, 4-1/2", Matscot Intl., 1990s	8.00	12.00
Mug, ceramic, "If you can read this you are an abductee," 1990s	8.00	12.00
Necklace, pewter pendant, "Believe," 2", no mark	5.00	8.00
Notepad, alien head, 75 sheets, 5-1/2", Color book, 1997	2.00	4.00
Panties, alien head graphic with "Believe" around the waistband, 1990s	5.00	10.00
Pen, silver figural alien, 6-1/2", made in Taiwan, 1990s	2.00	4.00
Pens, UFO pens, 4" x 7-1/2" package, Bic, 1990s	1.00	2.00
Pencils, alien pencils, 3" x 9" package with graphics, Dixon, 1990s	1.00	2.00
Poster, green UFO against blue stormy sky, "I Believe."	6.00	12.00
Poster, "Alien Backs Clinton," *Weekly World News* cover blow-up, 1990s	6.00	12.00
Poster, "Been There, Destroyed That," large green alien head with explosion in background	6.00	12.00
Poster, "Sightings," shows 18 UFO photos in color	6.00	12.00
Poster, huge alien head and hand over earth	6.00	12.00
Poster, "Take Me To Your Dealer," alien lights joint with flaming finger	6.00	12.00
Poster, "This is a Test of the Emergency Poster System..." with big civil defense alien head	6.00	12.00
Poster, "January 1, 2000—Full scale colonization begins," Western Graphics Corp., 1990s	6.00	12.00
Salt and pepper set, white figural alien heads, "for out-of-this-world taste," boxed, 1990s	7.00	10.00
Tie, silk, various alien graphics, 62", Renaissance, A. Rogers, 1990s, each	8.00	12.00
Welcome mat, "Aliens Welcome," coarse fiber with red and black print	15.00	25.00

Christmas stocking-$20.

Christmas ornaments, each-$10, MIP.

Diary-$8.

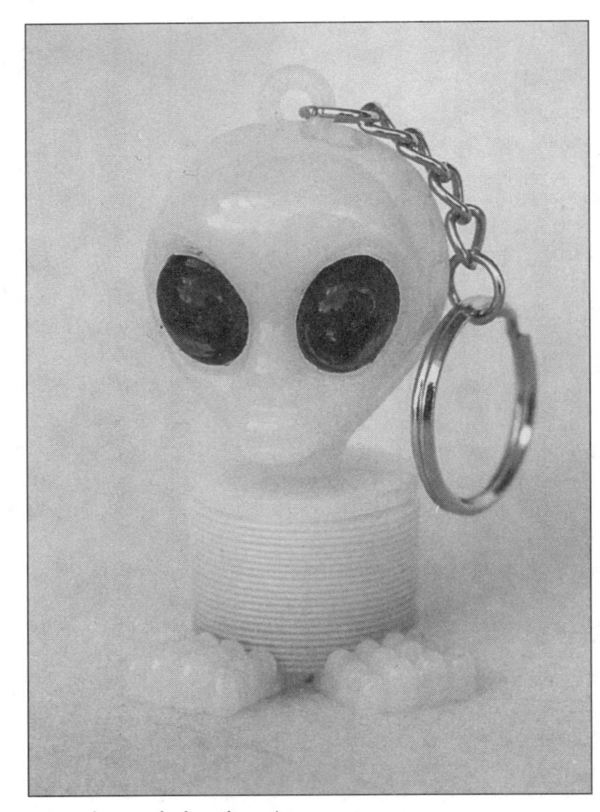

Key chain, slinky alien-$6.

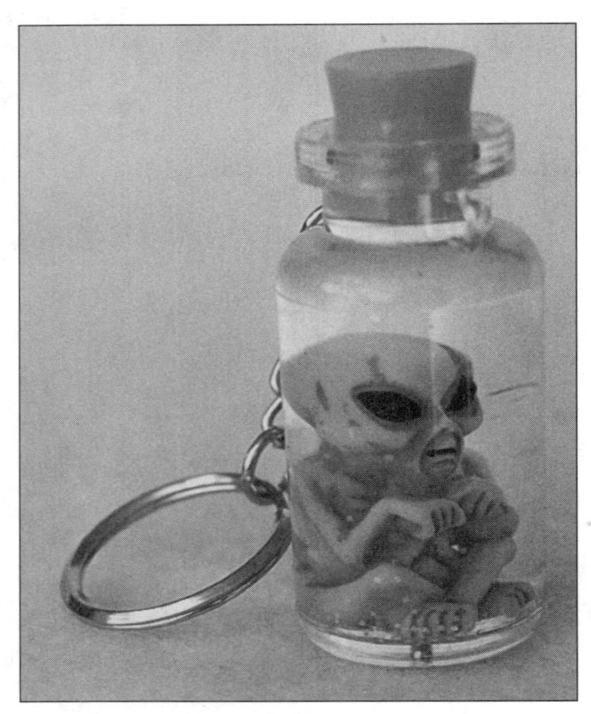

Key chain, "Image Bottle"-$4.

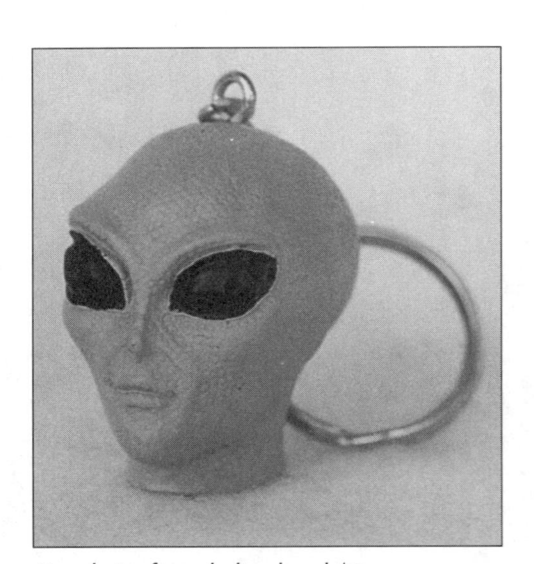

Key chain, figural alien head-$3.

Lamp-$35.

Mug-$12.

Necklace-$8.

Notepad-$4.

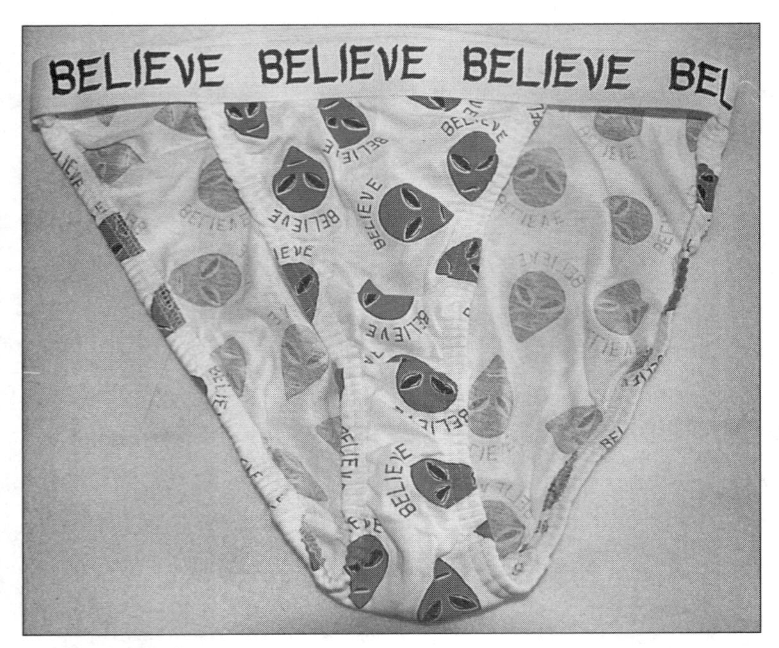

Panties-$10.

Pencils-$2.

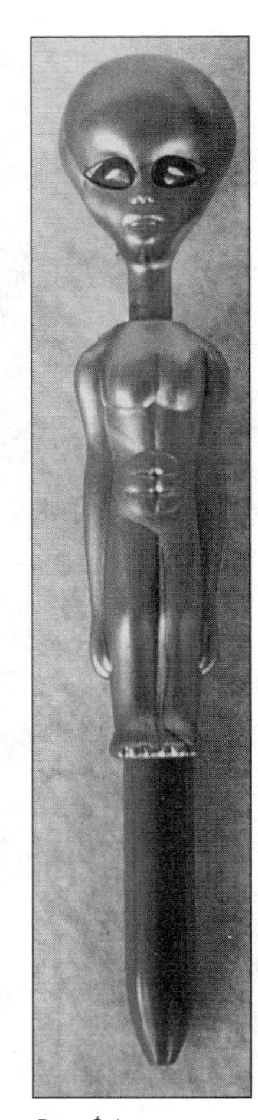

Pen-$4.

Pens-$2.

Poster-$12.

Poster-$12.

Poster-$12.

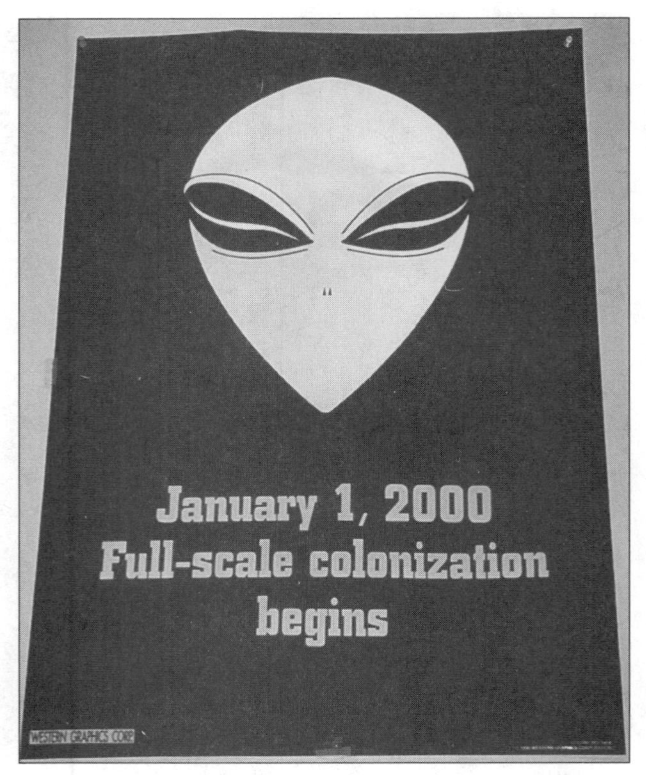

Poster-$12.

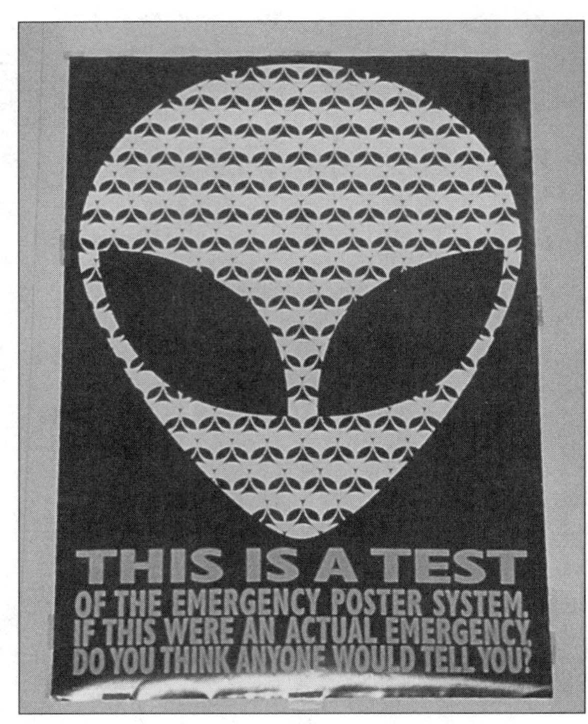

Poster-$12.

A trio of posters, each-$12.

Salt and pepper set-$10.

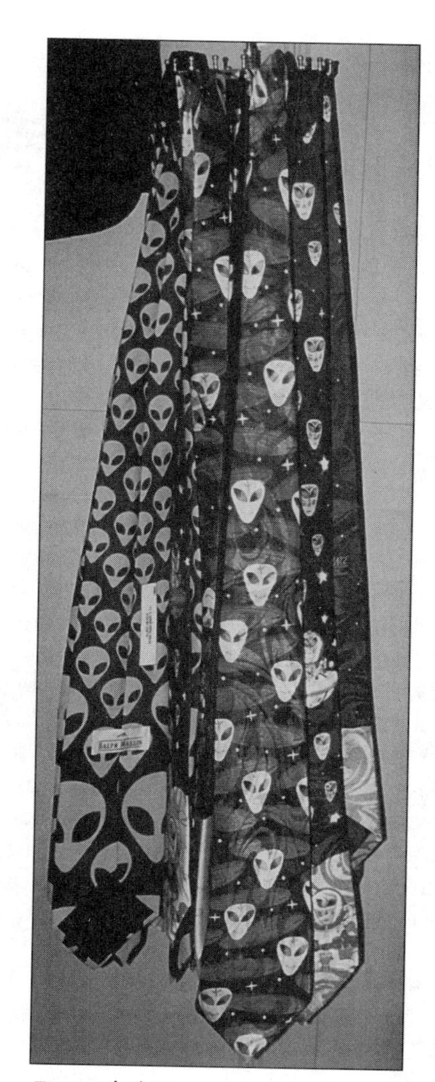

Tie, each-$12.

Welcome mat-$25.

Chapter 6
Roswell Mania and Area 51

In early July 1947, Roswell, New Mexico, became the center of perhaps the biggest UFO story in global history. Although actual dates vary (from June 24 to July 7), it was reported that a flying saucer crashed on a local ranch, and four aliens were recovered from the crash site. Two were dead, one died within a few hours, and one survived in top-secret captivity for about two years, according to some reports.

Government and military intervention reportedly silenced all of the locals who had come into contact with the saucer or its inhabitants. Soldiers at the Roswell Army Air Field were quickly transferred out, scattered to other bases around the country to prevent information circulation and to assure that if one of them talked, they would be pinpointed. Some witnesses report being threatened, while others were bribed. No one was to talk. Perhaps this is why information on "The Roswell Incident" has been so slow in coming out to the general public. After decades of struggling to get more information, it seems that at least some of the Roswell mystery has finally been revealed. In fact, the Roswell silence has given way to Roswell Mania. During the July 4 weekend in 1997, thousands of believers and UFO fans flocked to Roswell for a 50th Anniversary Festival to remember the legendary crash. The small town, nestled in the New Mexico desert in the southeastern corner of the state, exploded into a surrealistic UFO Mecca.

The event's mass popularity was fueled, in large part, by Fox Television. Fox's popular series *The X-Files* made the concept of earthbound aliens being concealed by the government a household notion. The network also broadcast a special made-for-TV film called *Roswell*, which reenacted much of the Roswell incident. And, who can forget the historic broadcast of "Alien Autopsy," a special narrated by *Star Trek*'s Jonathan Frakes, which supposedly showed the public actual footage of an autopsy performed on one of the recovered alien bodies from the Roswell crash in 1947. The film, uncovered by British rock promoter Ray Santilli, was aired on August 25, 1995, by Fox and a number of television stations around the world.

The "Alien Autopsy" special sparked much controversy and was even ridiculed by Fox's own *X-Files* characters in more than one episode. Many experts, however, believe the film to be authentic, despite the fact that the alien on the autopsy table does not match the descriptions of many of the Roswell eyewitnesses.

In the wake of the media hype, Roswell has now dubbed itself "The UFO Capitol of the World," and a trip to the downtown area supports the claim. The International UFO Museum & Research Center, at 114 N. Main St., is surrounded by a host of alien-themed gift and souvenir shops, and many local businesses sport signage directed at UFO fans.

For the collector of UFO and alien memorabilia, a trip to Roswell is a fun and essential pilgrimage. Here, you can find virtually every alien and UFO toy and gizmo on the market, along with booklets and 'zines by the eyewitnesses, many autographed, which offer more serious and rare firsthand accounts of what really happened in Roswell, New Mexico, in July 1947.

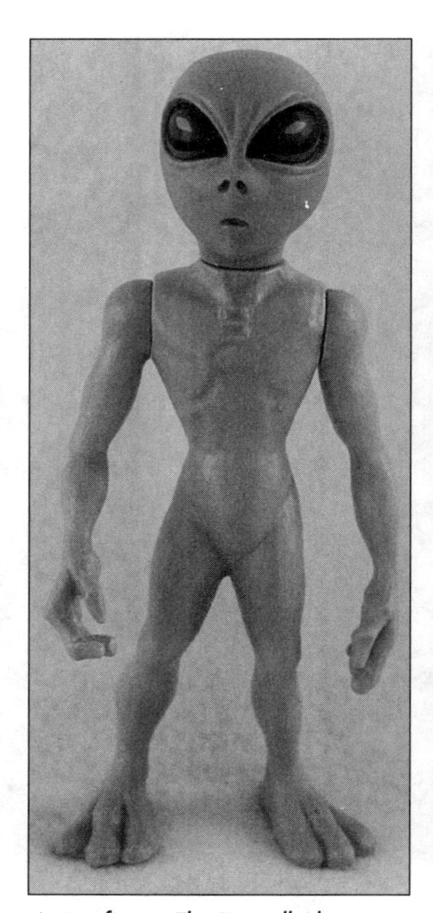

Action figure, The Roswell Aliens-$15, MIP.

Alien Pops container-$25.

Badge, Roswell RAAF Crash Recovery Team Access Pass-$5.

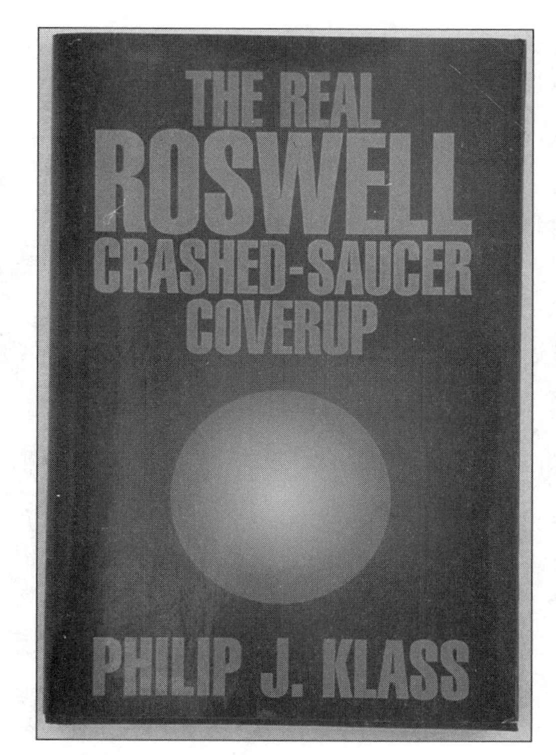

Book-$18.

Roswell Memorabilia

Item	Good/Loose	Mint/MIP
Action figure, The Roswell Aliens, 6", carded, Street Players Holding Corp., 1996	8.00	15.00
Alien Dissection Doll, with organs inside, green rubber, 14-1/2", R. Marino, WPF Inc., 1997	12.00	18.00
Alien Pops container, green plastic with removable sign on back, 1990s	15.00	25.00
Badge, Roswell Army Air Field Pass, 4" x 2-3/4" laminated badge, 1990s	3.00	5.00
Badge, Roswell RAAF Crash Recovery Team Access Pass, Hangar 84, laminated badge, 1990s	3.00	5.00
Book, *Real Roswell Crashed Saucer Cover-Up, The*, Philip Klass, Prometheus Books, 1997	12.00	18.00
Book, *Roswell Incident, The*, Berlitz and Moore, hardcover, 1st Ed, Grosset & Dunlap, 1980	10.00	15.00
Bumper sticker, "I crashed in Roswell," black ink, white sticker, 1990s	2.00	4.00
Bumper sticker, "Roswell, NM—A Great Place to Crash," black ink, white sticker, 1990s	2.00	4.00
Cap, "Alien Space Craft? UFO Research Center," 1990s	10.00	15.00
Cookies, Roswell UFO Incident Mini-Cookies, Santa Fe Cookie Co., 1997	3.00	7.00
Figure, Alien Autopsy figure, glows, 3-1/2", carded, Accouterments, 1996	1.00	3.00
Figure, Roswell Alien, 4", from *Aliens...The Mini Series*, Shadowbox Collectibles, 1990s	5.00	8.00

Bumper sticker, "I crashed in Roswell"-$4.

Cap-$15.

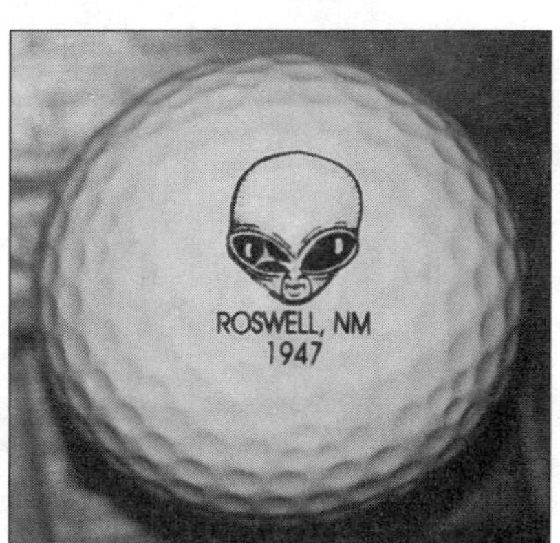

Golf ball-$5.

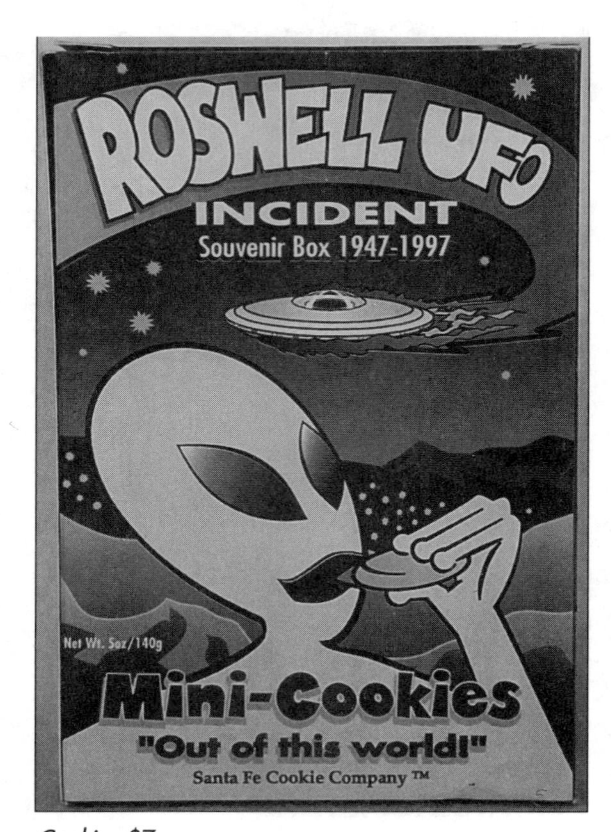

Cookies-$7.

Game-$15.

Item	Good/Loose	Mint/MIP
Game, Alien Autopsy Game (similar to "Operation"), DaMert Company, 1997	8.00	15.00
Golf Ball, small alien head graphic, "Roswell, NM"	3.00	5.00
Jewelry, space ship pendant on Roswell card, 50th Anniversary, Pfahler Mfg., 1997	10.00	15.00
Jewelry, alien head pendant on Roswell card, 50th Anniversary, Pfahler Mfg., 1997	10.00	15.00
License, Space Vehicle Operator's License, 3-1/2" x 2" laminated card with picture,1990s	2.00	5.00
License, Official UFO License, 3-1/2" x 2" laminated card with picture, 1990s	2.00	5.00
Model kit, The Roswell UFO, 1/48 scale, 7" long, 50th Anniversary, Testors, 1997	25.00	32.00
Mouse pad, Intl. UFO Museum & Research Center, 1990s	10.00	15.00
Mug, ceramic, "Alien Zone, Roswell, New Mexico," gift shop souvenir, 1998	4.00	8.00
Mug, Intl. UFO Museum & Research Center, ceramic, 1990s	5.00	8.00
Mug, "Roswell Daily Record" 1947 cover story reprint, ceramic, 1990s	5.00	8.00
Newspaper reprints of July 1947 Roswell Daily Record front pages with saucer stories (pair)	3.00	5.00
Poster, Roswell Fox TV movie poster, shows Kyle Maclachan and Martin Sheen, 1990s	20.00	30.00
Roswell Autopsy Alien, life-size display model, Michael Burnett Productions Inc., 1990s	350.00	450.00
Roswell Incident, 50th Anniversary Commemorative ltd. 2-piece display, Shadow Box, 1997	25.00	35.00
Tank top, "Roswell Encounter," with alien graphic in New Mexico symbol	12.00	18.00
Tapestry, "International UFO Museum," black and white blanket style with fringe, large size	20.00	40.00
Time magazine, cover story, The Roswell Files, June 23, 1997	2.00	5.00
T-shirt, alien holds big red heart that says "Roswell NM"	12.00	18.00
T-shirt, "Roswell NM is c-ALIEN-te," shows alien driving caliente pepper vehicle	12.00	18.00
T-shirt, small green alien head with "Roswell, NM "underneath	12.00	18.00
T-shirt, "Alie-ANN & ANDY" shows Raggedy Ann and Andy as aliens	12.00	18.00
T-shirt, "Alien Airbus of Roswell NM—Over 50 Years of Flying and Only One Accident"	12.00	18.00
T-shirt, "Most Wanted," with alien arrest graphic	12.00	18.00
T-shirt, "We've got Aliens coming out the Ying-Yang,"	12.00	18.00
T-shirt, alien postage stamp design with Roswell stamp	12.00	18.00
T-shirt, aliens in "hear no evil, see no evil, speak no evil" pose	12.00	18.00

Jewelry, space ship pendant-$15.

Jewelry, alien head pendant-$15.

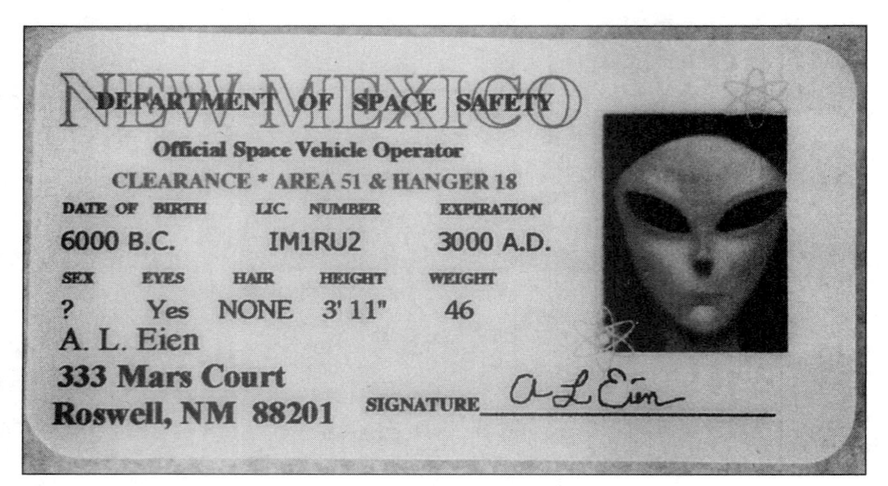

Space Vehicle Operator's License-$5.

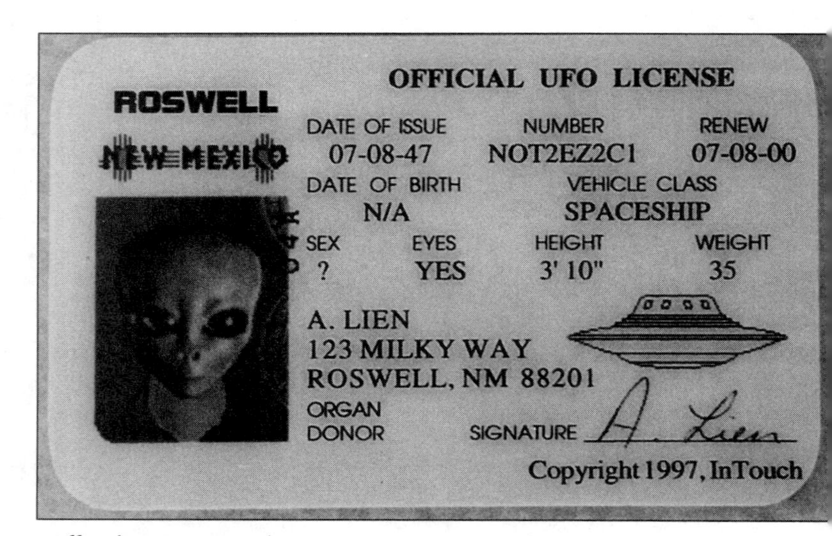

Official UFO License-$5.

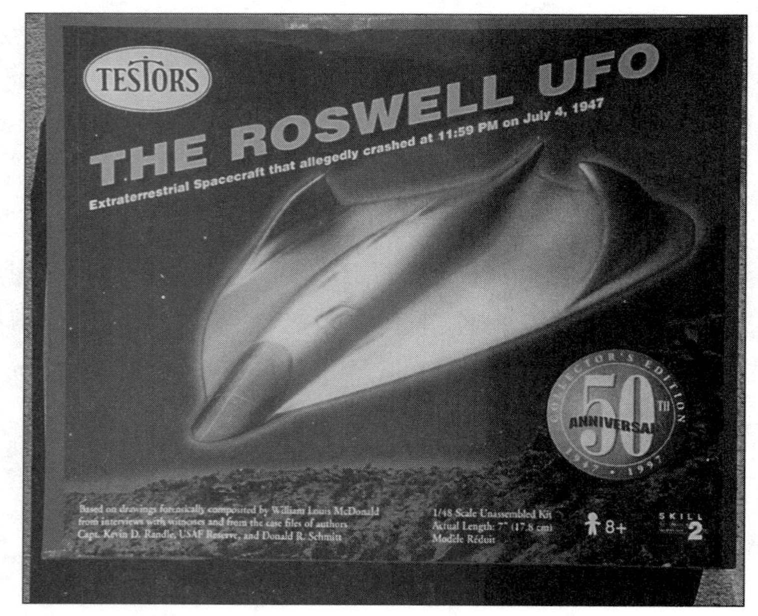

Model kit-$32.

Item	Good/Loose	Mint/MIP
T-shirt, alien on flying cycle, "Illegal Alien"	12.00	18.00
T-shirt, "Alien Spacecraft? UFO Research Center"	12.00	18.00
UFO Crash Site Souvenir Rocks	.50 cents	1.00
UFO magazine, cover story, Roswell Crash, Vol. 9, #2, 1994	3.00	6.00
UFO magazine, cover story, Roswell, Part II, Vol. 9, #4, 1994	3.00	6.00
UFO magazine, cover story, Roswell—50 Years and Counting, Vol. 12, #3, May/June 1997	3.00	6.00
UFO Universe magazine, cover story, Roswell Alien Autopsy Caught on Film, Winter 1996	3.00	6.00
Video Series, International UFO Museum & Research Center Lecture Series, available from the Museum, 1997-98		
Dennis Balthaser - What Are We Seeing?, Jan. 1997		15.00
Ted Loman - Crash Retrievals, Feb. 1997		15.00
Donald Schmitt - The Missing Nurse, March 1997		15.00
Clifford Stone - Operation Blue Fly, April 1997		15.00
Dennis Balthaser - A Brief Description of Area 51, May 1997		15.00
Donald R. Burleson - Piecing together the Roswell story, June 1997		15.00
Stanton Friedman - Flying Saucers ARE Real, Sept. 1997		15.00
Dennis Balthaser - Interception, Oct. 1997		15.00
David Davidson - The World's Mysterious Places/ Integration Time Machine, November 1997		15.00
Ellsworth LeBeau - UFOs and the American Indian, Dec. 1997		15.00
Deon Crosby - The Roswell Report: Case Closed, Jan. 1998		15.00
Wayne Mattson - History of White Sands Missile Range, Feb. 1998		15.00
Chet Sapalio - Possible Propulsion Systems of UFOs, March 1998		15.00
Clifford Stone - Plausible Deniability, April 1998		15.00

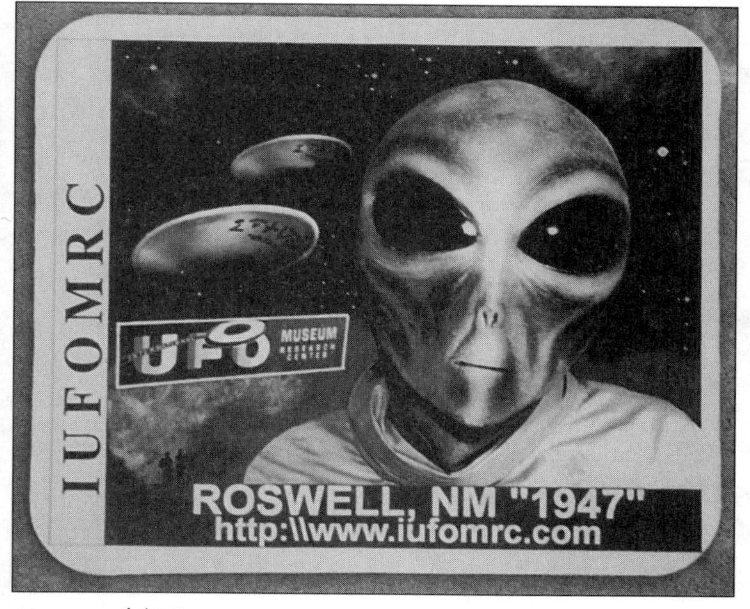

Mouse pad-$15.

Various mugs-$8.

Roswell Daily Record

Gen. Ramey Empties Roswell Saucer

Lewis Pushes Advantage in New Contract

Sheriff Wilcox Takes Leading Role in Excitement Over Report 'Saucer' Found

Arrest 2,000 In Athens in Commie Plot

Send First Roswell Wire Photos from Record Office

Ramey Says Excitement Is Not Justified

U. S. Land-Lease To Britain Looms As Needed by Fall

Local Weatherman Believes Disks to Be Bureau Devices

Romania Rejects Bid to Take Part in Economic Meet

Attorney to Force Closing up of Ruidoso Clubrooms

Find Nude Body of Strangled Woman in New York Hotel

Harassed Rancher who Located 'Saucer' Sorry He Told About It

Welcome to Roswell

Newspaper reprints of July 1947 Roswell Daily Record front pages-$5.

Poster-$30.

Roswell Incident, 50th Anniversary Commemorative ltd. 2-piece display, Shadow Box, 1997-$35.

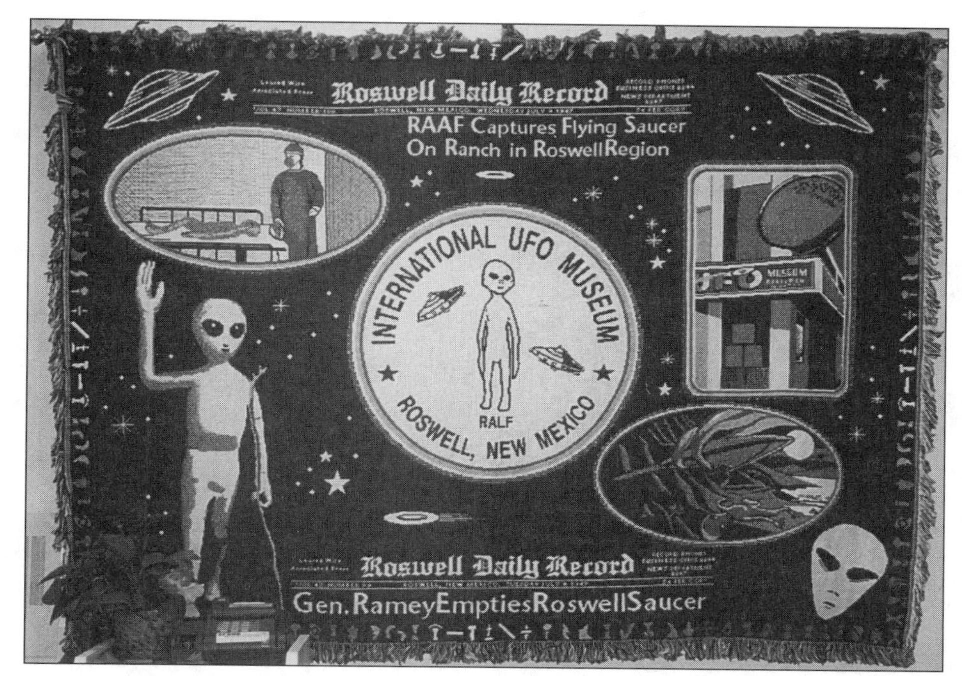

Tapestry-$40.

Tank top-$18.

Time magazine-$5.

UFO Crash Site Souvenir Rocks-$1.

T-shirts, each-$18.

Alien and UFO Collectors have a variety of T-shirt designs to choose from, such as these four; each-$18.

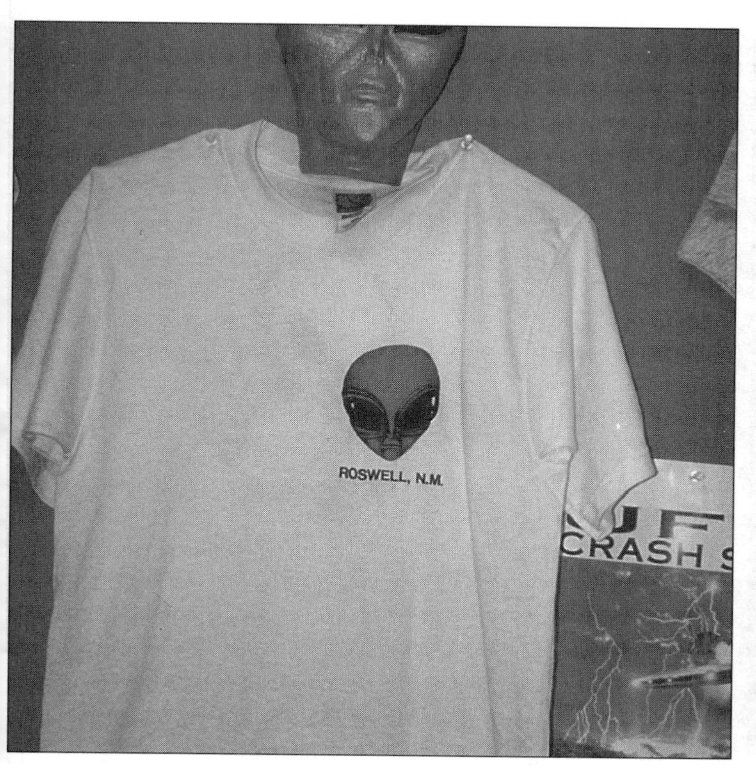

More T-shirts with various alien slogans; each-$18.

UFO magazine, Vol. 9, #2, 1994-$6.

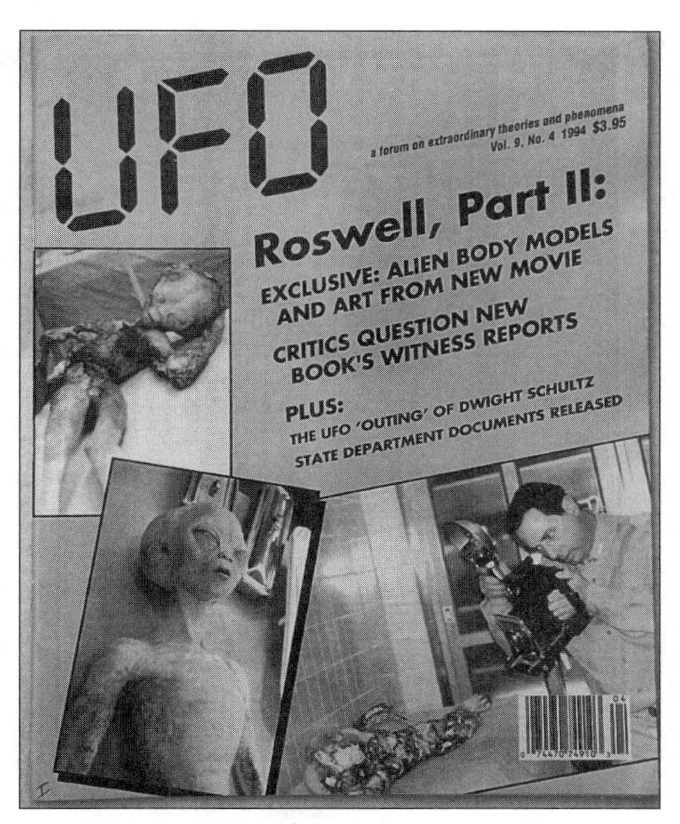

UFO magazine, Part II, Vol. 9, #4-$6.

UFO magazine, Vol. 12, #3-$6.

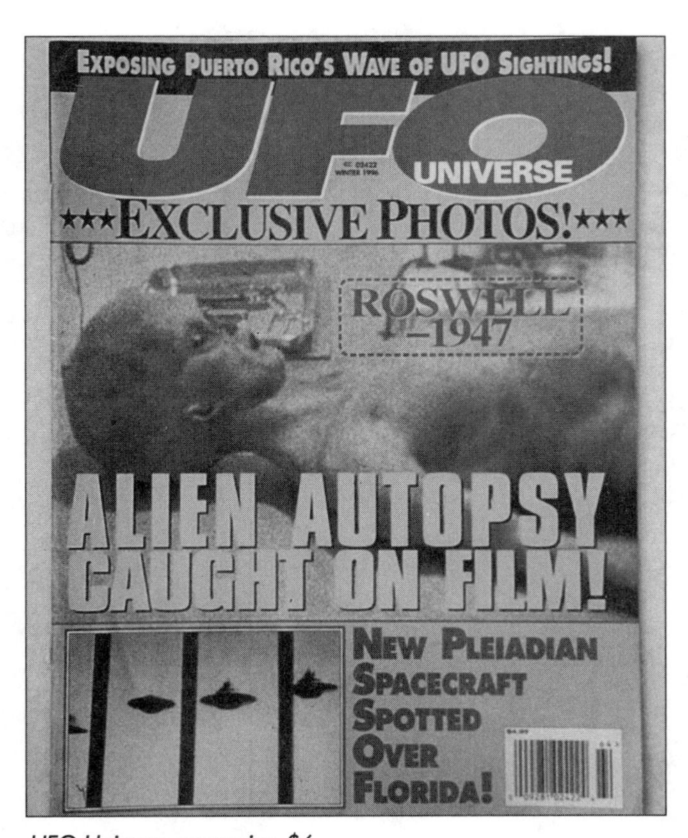

UFO Universe magazine-$6.

Item	Good/Loose	Mint/MIP
Donald R. Schmitt - Was There a Survivor at the Roswell UFO Crash?, May 1998		15.00
Paula Rich-Greenwood - The Extraterrestrial Phenomenon: Past, Present & Future, June 1998		15.00
Norman Melvin - When Truth Hides in Fiction, Aug. 1998		15.00
Dennis Balthasar - Underground Bases, Sept. 1998		15.00
Clifford Stone - Air Force Confirmation - Interplanetary Spacecraft, October 1998		15.00
Other videos available from the museum:		
Another Look at the Jim Ragsdale Story, Dennis Balthaser, MUFON Meeting, March 1997		15.00
Roswell Incident: 50th Anniversary Witness Round Table, July 1997		15.00
Founding Members Banquet with Bob Barnes as "Mac Brazel Resurrected," Dec. 1997		15.00

Badge-$5.

Area 51

The nation's other notorious UFO hot spot, Area 51, is in a remote region of Nevada. Area 51 is a heavily-guarded government base, where many believe alien space ships—and possibly alien beings—are being held and studied by our military.

Item	Good/Loose	Mint/MIP
Alien in a jar, tagged "Area 51," Astonishing Collectibles, 1990s	20.00	30.00
Badge, Area 51 Special Agent badge, 4" x 2-3/4" laminated badge, 1990s	3.00	5.00
Cap, Area 51 logo	10.00	15.00
Model kit, Area 51 UFO, 13 in. diameter, 1/48 scale, boxed, Testors Corp., 1996	25.00	32.00
Pamphlet, Area 51 and S-4 Handbook, by Clark, from IUFOMRC	15.00	20.00
T-shirt, Area 51, safari green, M, L or XL	15.00	20.00
T-shirt, Area 51 - Restricted, M, L or XL	20.00	25.00
T-shirt, Area 51 - Secret Facility, M, L, XL or XXL	15.00	20.00
T-shirt, "Danger, Top Secret, Area 51 - I breached AREA 51 - If I tell you what I saw ..."	12.00	18.00
Video, Secrets of Dreamland II, from IUFOMRC	30.00	38.00

*** Note: IUFOMRC is the International UFO Museum & Research Center in Roswell, New Mexico.**

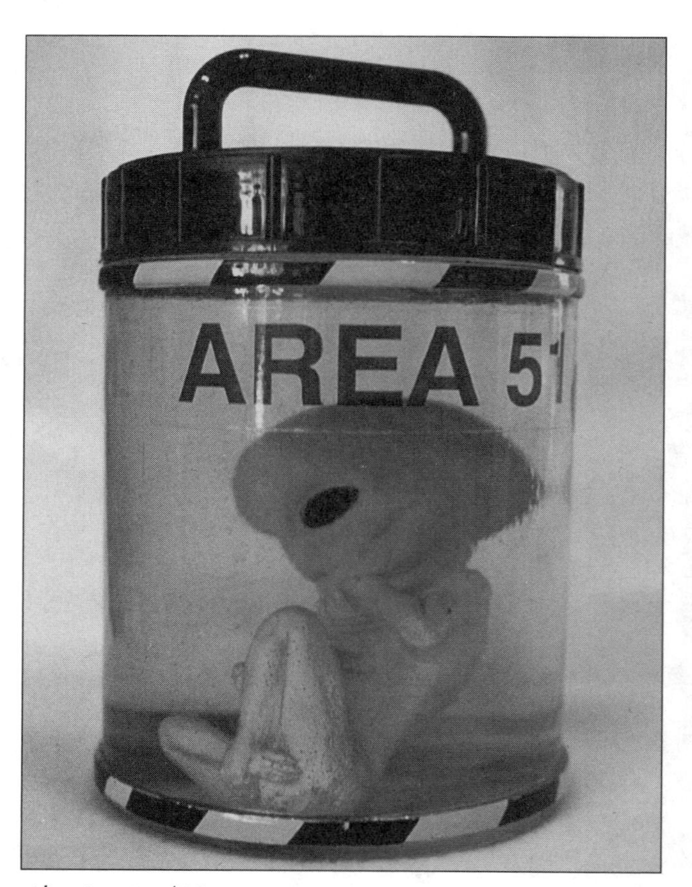

Alien in a jar-$30.

Model kit-$32.

T-shirt-$18.